# ★ OUT OF THE TRENCHES ★

## LESSONS FROM THE TECHNOLOGY FRONT LINE

Advice from a real estate software mercenary to help take the fear and risk out of your migration from one major software platform to another

## By David Wolfe

# *Testimonials*

*"David Wolfe's straightforward approach to software implementation proves he is a master in his field. He shares his vast knowledge of the intricacies unique to the real estate industry and charges us to leave our egos at the door. In his over thirty years of experience, he's seen and heard it all, from unnecessary interruptions and delays to slipshod training. In a matter-of-fact style, David's methodical and sound advice will benefit every level of employee, from the savvy professional to the most tech-challenged individual.*

*In his first book, 'Software and Vendors and Requirements, Oh My!,' David provided a step-by-step guide to choosing software. For companies that have spent valuable resources and money on choosing new software, the work doesn't stop there. The cost of failure for a conversion is high— and unnecessary. A company's success depends not only on its expertise in real estate, but also on its willingness to understand its limitations. Migrating from one software system to another is something David Wolfe does every day. Read his book and then hire the expert."*

Joan Mora
*Chief Financial Officer*
*Tonti Properties*

---

*"One of the most frustrating exercises in business can be dealing with systems and software consultants. Those of us who've sat across the conference room table from an expensive team of outsiders know how painfully opaque such projects can seem. And how expensive—and threatening to your own survival—these projects might be if they blow up. David Wolfe has generously turned his vast experience on the other side of the table into plainspoken transparency. His advice in this book provides a clear map to navigating these waters knowledgeably and*

*safely. I can't imagine diving into such a project and trying, without this handy and often humorous manual, to keep all the relevant questions and issues in my head. Whether you're a top executive who needs to understand what's about to happen in a software-conversion project, or you're the poor sap who must manage such executives' expectations, 'Lessons from the Technology Front Line' will save you enormous confusion and headaches."*

Mark Obbie
*Writer and Editor*

———

*"When David asked if I would write a testimonial for his forthcoming book on software migrations, I immediately jumped at the opportunity. Lupine Partners has proven to be an invaluable resource for our company—a large multifamily mixed portfolio property manager and developer out of Chicago—not only in planning and preparing for our migration from MRI to Yardi, but also in project managing the whole conversion process for us.*

*Before hiring David and co., we did our due diligence on and interviewed several prominent implementers. It rapidly became clear to us that Lupine had developed and followed a specific methodology to software conversions that would greatly benefit us if properly applied. The implementation and conversion philosophy outlined in this book follow David's style of clear, no-nonsense and direct communication that we have seen in action every day of our conversion—David and his team practice what they preach!*

*I highly recommend that anyone looking to go through a major software conversion or implementation of any kind buy, read and thoroughly assimilate this book. The thought-provoking lessons and concepts conveyed will serve as a road map for you as you seek out implementation partners and approach your own implementation and conversion projects."*

Mark Sweeting
*IT Director*
*Draper and Kramer*

*"Real estate enterprises have undergone tremendous increases in both size and complexity over the last twenty years; during this same time period, real estate enterprise software has undergone significant evolution. As a result, business leaders who wish to remain competitive are likely to face the daunting prospect of 'swapping' one software platform for another at some point during their careers. In 'Lessons from the Technology Front Line,' Wolfe provides a clear, much-needed road map to navigating this tricky process."*

Anthony Page
*Senior Vice President & Director*
*Capstead Mortgage Corporation*

---

*"I have been involved in my share of software projects—most have gone very well. And some—well, not so much...*

*I read the excerpted chapters of David Wolfe's latest book with interest (particularly since I am going through another software migration at this very moment...). I think Wolfe has done a very good job of describing how to break the implementation effort into manageable, concurrent steps. His approach makes sense: Plan, orchestrate, execute. His analogy to building a house is spot on—get a roof over your head for that first. Work on the front lawn after you move in.*

*This is a timely and good book for anybody thinking about changing to a new software system."*

Carl Pearce
*Controller*
*Pillar Commercial*

*"I read with interest David Wolfe's new book, 'Lessons from the Technology Front Line.' Although a veteran professional in the multi-family industry, I have not been through many software conversions. I am going through one right now... Mr. Wolfe gives solid, no-nonsense advice. His battle scars are obvious. It is also obvious that he honestly wants to keep his readers from making the same mistakes that others have made. I think this is a terrific book to read for anybody who is going through, or planning to go through, a change in software."*

Cindy Ash
*President*
*Embrey Management Services*

———

*"I have known David Wolfe for 40 years now—back before both of us were shaving…He has been in the software business for as long as I can remember. But until he asked me to review his latest book, I didn't quite know exactly what he did.*

*Now I do.*

*I hate to admit this to him, but I am impressed with what he has written, and his writing talent. And the more I read, the more I realize that his experience is something I need at my company as we are going through a major ERP software upgrade. I am passing Wolfie's book on to my IT staff—they may have some splainin' to do. Nice work, old friend."*

Dan McCutcheon
*President*
*Star Pipe Products*

*"I recently finished a software implementation for my firm. Admittedly, there were some challenges that I needed to overcome to manage the process and get the implementation done correctly upfront. I would have loved to have read this book prior to getting started on our project. The book exposes you to the terminology needed to communicate with IT professionals, thoughtful process overview of an implementation and strong management tools to manage your team and the interaction with consultants. David Wolfe has provided a terrific road map on how to get from here to there. His grasp of the subject matter is excellent and the information in the book is presented in a thoughtful manner."*

David DiSanto
*Trilogy Real Estate Group*

———

*"My family has owned multi-family real estate investments for over 50 years now. We have always been committed to using technology strategically to give us a competitive advantage. While I am not a computer guy, I am interested in it and I am interested in the application of technology when it makes sense for my company.*

*David Wolfe's book does an outstanding job of outlining the risk points when moving to a new software system. He speaks in a plain voice and does not try to confuse the reader using fancy computer language. He tells it like it is—the good, the bad, and the ugly."*

Don Pardue
*President*
*Pardue Investments*

*"David Wolfe's chapter (ten) on training makes the whole book worth it. I have been an executive in real estate operations for over 12 years. I have been involved in many, many property takeovers, and more than my share of software conversion efforts. To me, the biggest success/failure point with these conversion efforts is this: Can the on-site people use the new software system proficiently?*

*I have seen some good training sessions over the years, and I have seen some really bad ones. Bad training=bad implementation. And vice versa. Anyone who is thinking of moving to a new software package would be well advised to read this book."*

> Donna Summers
> *Vice President*
> *Gables Residential*

———

*"Having recently completed the first phase of a major global software implementation, I have experienced the highs and lows based on several key points made in David Wolfe's latest book. Smart planning ahead of execution is very important in any size project and it gives the user community top-notch training before execution. It not only provides a successful transition between the old system and the new system but also helps engage the user community and provide support for the project in its entirety.*

*The author clearly emphasizes how given good implementation techniques avoid compromising a project. His common-sense approach on software implementations is easy to follow and provides so many opportunities to make the right choices before, during, and after the implementation.*

*This book definitely goes on my bookshelf in my office."*

> Laura Gill
> *Senior Business Analyst*
> *ProLogis*

*"I am a real estate professional—most definitely NOT a computer guy. I don't know a bit from a bite (byte?). When David Wolfe asked me to review his book I groaned inside. There goes my weekend...*

*Then I read how he got started and how he worked for John Connally many years ago. And then read some more and realized that his approach (and the entire book) didn't really have that much to do with computers. What he really seems to do is manage fear, and keep people from acting against their own best interests. This book has a parental feel to it. I can tell he is a stern and loving parent to his client children.*

*This was a pleasant surprise."*

Gordon Dey
*Broker/Associate*
*Keller Williams Realty Gulf Coast*

--------

*"This book is a must-read for real estate accounting and IT professionals who are planning to migrate to a new property management software solution or module. The author, David Wolfe, is a CPA who draws upon his two decades of experience as a software consultant in the real estate industry to provide an easy-to-read, practical approach to planning and executing a successful implementation. The process of migrating to a new software platform is a challenging process that is not customary to the users at any level, and it is generally an additional task to everyday business operations. This book provides guidance for how to accomplish a migration successfully from the inception of identifying good candidates for the project team and project manager through the go-live date. The book provides step-by-step instructions on how which tasks need to be accomplished and at what point, useful tasks that a team of users may not think of on their own. By following the advice in this book, one could avoid the problems that plague unsuccessful implementations."*

Jennifer van Arcken
*IT Director*
*BSR Trust*

*"Having read the first few chapters of David Wolfe's book, 'Lessons from the Technology Front Line,' it is obvious that he has a command of software implementation tactics—and that he has the requisite experience to help his readers avoid common (and not so common) pitfalls. This is an enjoyable read. Besides being an accomplished technology consultant, Wolfe has some writing chops as well. This is no technical tome…*

*My congratulations to the author."*

> Jorge Cordova
> *President*
> *Synterna Technologies*

———

*"In 'Lessons From the Technology Front Line,' David Wolfe brings a much-needed dose of 'been there, done that' reality to delivering a successful software implementation. As an IT Project Manager myself, I've lived through the very challenges of people, process, data, and technology that David speaks about in this book. He meets these challenges head-on with practical advice, tools, and templates your organization can use right now. Hard-earned lessons David has acquired over many years in this business come to life in the personal examples and anecdotes he includes throughout. If you need some straight talk about what to emphasize and what to avoid in your software project, 'Front Line' is an outstanding read!"*

> Keith Richey
> *Vice President*
> *Propoint Technology*

*"Several years ago I had the pleasure of working side by side with David once during a software implementation and another time as part of a company's software selection process. What David recommends in his book is the same methodology he uses to run these engagements. His consistency and faithfulness to his approach is absolutely right on the mark. He asks critical questions and leads the team to think about and consider people, situations, and assumptions that if not given appropriate energy, can completely derail the entire process. If you are starting the process of moving to a new software system, talk to David and his team. And more importantly, read this book. You will save yourself time and money, and ensure a successful implementation."*

Kendall Pretzer
*President*
*The Strategic Solution*

---

*"Where was this book last year when we went through two simultaneous software implementation projects? David Wolfe has written a thorough instruction manual on how to navigate the move to a new software system. It is easy to read and contains real-life examples of how to approach these projects. His experience with these conversions is evident.*

*David's experience, and rational, systematic approach to this topic makes for an informative read. I highly recommend 'Lessons From the Technology Front Line.'"*

Lori Creasy
*Chief Financial Officer*
*Washington Nationals*

*"I was fairly convinced that hiring a consultant to assist us with our conversion would only result in higher costs and a watered-down result. Then my president and I met with David Wolfe. Talk about a 180-degree turnaround in perception! David's initial presentation to us was so concise, well thought-out, and thorough, that I had no doubt that we should hire him and his company to help us with our conversion. I couldn't have been more pleased with our results. David's approach and scheduling techniques set a clear path and task assignments through every step of the process. The finale to our project was three separate conversions over a 2-month period: all occurred on schedule and all were successful. His staff worked very well with us, encouraged us, held our hand when we needed it, and made what seemed a daunting task at first occur in bite-sized pieces until our plate was empty. Will we hire Lupine Partners again? We already have on several smaller projects. They have become a go-to partner for our company, whether we're needing a customized report, or further expanding our utilization of the software's capabilities. I highly recommend David Wolfe and Lupine Partners to anyone considering a software conversion."*

Joanne Massey
*Controller*
*Sundance Square*

---

*"I have worked with Lupine Partners and David Wolfe on multiple software implementations. David's methodology and approach to implementing are easy to understand, simple to follow, laid out in a logical matter and useful when applied. His process helped me build a long-term, lasting and successful software implementation and training team at my company. This success drove applying his methodology on projects outside of the engagement with Lupine."*

Matt Kraft
*Vice President of Training*
*Associa*

*"I had the privilege of working with David Wolfe on his first major software implementation in 1994. We served as co-project managers. What is interesting to me, and perhaps instructive to the readers of this book, is how much he has grown and matured in his approach to software implementation—but, more importantly, how of the core of his approach was there at the beginning of his career as a software professional.*

*Software is such an integral part of how all of us do our business now. Getting the wrong package, or implementing the correct package poorly, will place you at a competitive disadvantage quickly. I can vouch first hand that David Wolfe knows his stuff. And his approach will work for you. Just do what he says…"*

Michele Wheeler
*President*
*Jackson Shaw*

———

*"Having recently directed a major software conversion for our company I have only two things to say about David Wolfe's new book: 1) I can relate! and 2) I wish David's book had been available before our adventure. It's a must-read for anyone about to embark on the implementation of new operational software.*

*My only regret is not having access to David's new book before undertaking our recent software conversion!*

*David does what few software consultants do: he de-mystifies the process of software implementation so that even the most tech-challenged executive can understand it."*

Paul Votto
*Chief Operating Officer*
*Wimmer Brothers*

*"Save yourself countless hours of asking 'Where did we go wrong?' and start by reading David's book 'Lessons from the Technology Front Line.' This book guides you each step of the way on how to build your conversion team and make the right choices to make your implementation a success. There's much more to the project success then what reports the software generates. Without the right planning process you're likely to fail. 'Front Line' will give you the wisdom of hundreds of successful conversions, tell you about mistakes before you make them, and help you ensure that your project becomes a moment of triumph for your organization, one that helps you leverage your investment in the software."*

Phillip Jackson
*IT Director*
*Price Edwards and Company*

———

*"I have been in the software business for 14 years, and I have seen my share of botched software implementations. The principles and methodology that David Wolfe promotes in his software book are spot-on. So many companies try to do this on their own and fail—because they don't have a plan and, in many cases, they don't know what they are doing.*

*I endorse this book. Good work, David."*

Rick Greco
*National Sales Director*
*RETransform*

*"'Lessons From the Front Line' is a dead-on description of David Wolfe's implementation methodology. I have had the pleasure of having his firm lead us through a software migration. His book is the real deal and is exactly the approach he took in leading us through the process. His book is no college textbook, but a real-life, practical approach to going through a software project.*

*I highly recommend this book to anybody who is thinking about changing software packages. It is a small price to pay for peace of mind."*

Robin Brankamp
*Controller*
*Neyer Management*

———

*"As a business owner, I sometimes feel I am held hostage by the ever-changing nature of technology. I'm not saying I don't like software, but frankly, at times, it scares the hell out of me. I have been contemplating a migration to a new software product. My construction company has been using the same old package for a long time. I feel that, maybe, we are not working as intelligently as we could be.*

*But how to go about this change?*

*Then I ran across David Wolfe's book, 'Lessons from the Technology Front Line.' His book has supplied me with a step-by-step method for getting this process done. And he does it with no BS. I even laughed a few times while reading it. Pleasantly surprised.*

*This is a book that every business professional should read."*

Joey Gardner
*President*
*Simpson and Gardner*

*"David Wolfe has written a complete, concise and easy-to-read guide to implementing software. It is obvious that David is a veteran of many software implementation battles—and he has the 'been there, done that' experience to prove it. David is generous in his advice on what to do, and maybe even more importantly, on what NOT to do.*

*Too bad this book wasn't written a few years ago; I could have used it as a blueprint for some of my projects. Well done!"*

Steven A. Abney
*Chief Financial Officer*
*Prescott Realty Group*

———

*"I have worked with David on software projects since 1994—while employed at several different organizations and working with several different software packages. There is nothing theoretical about what David Wolfe writes about. His book gives good, hard-earned, and practical advice on an initiative that can be gut-churning. He has synthesized the process into something that looks like a software implementation blueprint...which I guess was his goal in writing the book. We go back a long way—he's the real deal."*

Steve Otermat
*Controller*
*Martin Fein Interests*

*"This is a good book.*

*In a previous life I performed these exact same services that David describes. My favorite part of the book is when he describes the choreography of the entire software migration process. He describes it as a 'dance' and to an extent it is. I never quite thought of it that way, but it is true. David has the art of software implementation down to, well, a science..."*

Sue Kilpatrick
*Software Implementation Specialist*

———

*"David Wolfe is an exceptional entrepreneur and a leader in the real estate software industry. David's book addresses the problems that many real estate professionals encounter when they try to utilize software on their own and provides the best practices in order to overcome these challenges. I highly recommend this book to any real estate professional who is seeking to better his or her business."*

Tom Bukacek
*Author of "Long Term Greedy" and*
*"The Real Estate Millionaires' Blueprint"*

*"Once again, David Wolfe hits it out of the park with his latest effort, 'Lessons from the Technology Front Line.' With his familiar honest, clear and pragmatic delivery, David provides a succinct and uncluttered roadmap to the software conversion/implementation process. In a style all his own, David walks the reader through a process he and his partners have executed hundreds of times. With a sprinkling of anecdotes and lessons picked up over a lifetime of effort, everything you'll need to know to achieve success (and avoid the pitfalls) in this complex and fragile exercise are made available. This book will stand proudly next to my copy of 'Software & Vendors & Requirements, Oh My!' as an indispensable resource and 'user manual' for the process of enterprise-level software migration. Anyone contemplating such a project (whether it's your first or your thousandth conversion) should read this book first—there's something here for everyone."*

D. Thomas Figert
*Chief Information Officer*
*Harbor Group International*

---

*"In chapter one the author says, 'Implementing software is not hard. The key is in knowing that the required steps must occur in a certain order though. Step C must follow step B which must follow step A.' It's a simple concept and, in my opinion, it is dead-on accurate. Putting it into practice is sometimes harder than it looks...*

*David Wolfe's entire book is geared towards telling the reader what needs to happen first, then next, then next...I think it is a terrific software 'cookbook.'"*

Wayne Parnell
*Controller*
*Lincoln Harris CSG*

*"I have read about some colossal systems failures over the years—ones where companies invest millions and millions of dollars with nothing to show for it. As I read David Wolfe's book on software implementation practices it seemed to me that many of these disasters could have been avoided just by following the rational and systematic methods that he advocates.*

*This is an important book and an easy read."*

Andrew Rottner
*President*
*North Texas Bank*

Published by CelebrityPress™, Orlando, FL
A division of The Celebrity Branding Agency®

Celebrity Branding® is a registered trademark
Printed in the United States of America.

ISBN: 978-0-9857143-4-5
LCCN: 2012948256

This publication is designed to provide accurate and authoritative information with regard to the subject matter covered. It is sold with the understanding that the publisher is not engaged in rendering legal, accounting, or other professional advice. If legal advice or other expert assistance is required, the services of a competent professional should be sought. The opinions expressed by the authors in this book are not endorsed by CelebrityPress™ and are the sole responsibility of the author rendering the opinion.

Most CelebrityPress™ titles are available at special quantity discounts for bulk purchases for sales promotions, premiums, fundraising, and educational use. Special versions or book excerpts can also be created to fit specific needs.

For more information, please write:

CelebrityPress™
520 N. Orlando Ave, #2
Winter Park, FL 32789
or call 1.877.261.4930

Visit us online at www.CelebrityPressPublishing.com

*For Tooduh*

# Contents

# A NOTE FROM
# THE AUTHOR

Throughout this book, you will see references to a "go-live date." All of the examples and hypotheticals in the book assume a cutoff date of Sept. 28 and a go-live date of Oct. 1. Therefore, the "dark" days are Sept. 29th and 30th.

You will also see references to Yardi, Yardi Systems, Yardi Voyager, and Voyager. Yardi and Yardi Systems refer to the real estate software company headquartered in Santa Barbara, California. Yardi Voyager and Voyager refer to the core software product of Yardi Systems. Yardi Systems, Inc. has not collaborated with me on this book, and none of the principles I outline should be considered their endorsement of our consulting approach or of the general content of the book.

I have included the references to Yardi for two reasons: One, it is the software product to which Lupine Partners has migrated the greatest number of clients over the past 20 years. It's comfortable for me to use this language. Two, I liked it better than saying "destination system" or "the software to which you are converting."

The principles outlined in this book can be applied to any software product being implemented in any industry.

# CHAPTER 1

# GETTING STARTED – YOU AND ME

"So what's my tale, nightingale?"

Back in the spring of 1982, I was driving back to Houston from a golf tournament my employer had sponsored. It was a Friday night and the traffic was really bad. As I was sitting there in the freeway parking lot, I told myself that I really missed Austin, where I had graduated from the University of Texas a few years earlier. A few months later, I answered a small two-line ad in the Houston Chronicle. An Austin-based real estate company was looking for a controller. What I didn't know was that the company was the Barnes-Connally Partnership, which was owned by former Texas Speaker of the House (and Lieutenant Governor) Ben Barnes and former Governor of Texas John Connally.

I interviewed, made the cut and got the job. I was 24 years old.

On my first day on the job, I got to meet Ben Barnes. I was ushered into his office to discuss my background. He asked me how old I was, and I told him. Ben, who always had a big fat cigar in his mouth, leaned over the desk and said to me,

"When I was 24, I had people who would kill for me." Well, okay…

A few months later, I had my first sit down with the governor. By then, I had power of attorney to sign loans on behalf of both men, and as the proverbial messenger, I was coming in to give Connally bad news about loans coming due and other unpleasant information. A lot of the meeting is a blur, because I was scared out of my mind, but what I do remember was something he told me he'd learned from Lyndon B. Johnson (to whom he served as an assistant in the 1940s). And it was this:

1. Answer your own phone.
2. Return all of your phone calls.
3. Speak plainly. Do not use fancy words or hide behind credentials.
4. Protect those who employ you.

I was fortunate to work there for two reasons. One, being around these two impressive men gave me a model of how to act and how to "be" professionally. As I sit here at my conference table this morning writing this chapter, I am 54 years old. Some 30 years later, I still find myself still going back to sessions with Mr. Barnes and Mr. Connally. These values are stuck in the back of my head and I frequently call on the lessons I learned there. Two, it was the beginning of the PC era, and I was fortunate to get my start then. In my role as controller, one of my jobs was to implement and configure the software to the best of my ability. In the 1980s and into the early 1990s, everything ran off of the DOS operating system. The tools were so limited then that in many ways, you actually had to be more clever to effect solutions. I was an okay accountant, but I leaned towards software, and was always very creative on that side of the business.

## IMPLEMENTATION METHODOLOGY

After the well-publicized Barnes-Connally bankruptcy in the late 1980s, I went to go work for a bank in Dallas. I served in the MIS department and supported the real estate operations group. I didn't know it at the time, but it was the beginning of what would become Lupine Partners. Before I left to start Lupine, though, the bank went through an implementation effort similar to the one you are contemplating.

It was my first software implementation effort, and it was three years before I founded Lupine Partners. I've learned a lot over the years about software implementation: a lot of what to do, and (possibly more importantly) I've learned what not to do. When I started Lupine Partners my hair was brown. It's now completely gray (some might even call it white). If you do anything in a leadership role for nearly twenty years, you begin to develop a methodology. At Lupine Partners, we have absolutely done that, particularly with regard to Yardi Voyager.

We have a clearly defined way of implementing Yardi Voyager, one that actually has its roots in the late 80s and early 90s, the period in which I was learning how to become a project manager. My boss at the time, Walter, had considerably more experience than I did. He gave me good advice, most of which I didn't listen to because I was thicker than a five-dollar malt. Then I got hung up on a few projects that I was handling in the "old David" way, and things didn't turn out so well for me. So I went to him, told him that he was right and I was wrong, and from that point on I really started building my own methodology. I used some of Walter's techniques and made them my own. I removed a number of tactics that I didn't want to use because they didn't fit my style, weren't relevant, or didn't even make sense to me. Over the years--going on twenty now--I still continue to add to, subtract from, and tailor this approach so that it works for Lupine and our clients. When we started working with Yardi about twelve years ago, we made some tweaks by adding tasks that recognized Yardi's role in the implementation process.

## OUR RELATIONSHIP WITH YARDI SYSTEMS

Lupine has been working with Yardi since 2000, when we were named project manager of the FirstWorthing implementation in Dallas, Texas. As of this writing (May 2012), we have provided services to 163 Yardi clients and have served as project managers, taking the client from initiation to final go-live, on 72 occasions. Along the way, we have developed a methodology that has been tweaked, modified, stress-tested, and refined. It works, it has worked, and will continue to work into the future. It is not the only way to implement Yardi Voyager. But it's our way and, in our opinion, the most direct route to get you to where you need to be.

Yardi Systems has been a terrific company with which to partner. They support their proven independent consultants with special conferences, access to their knowledge base, use of the software, and amazing support as we work together to lead our mutual clients through the process from software platform X to Yardi Voyager. Yardi is comfortable with our methodology and they continue to be an amazing partner. One of the best decisions I made as President of Lupine Partners was investing time in building a relationship with this fine organization.

## THE COST OF FAILURE

If you are in the market to implement Yardi Voyager, your main concern should be the cost of failure– of **repeatedly having to go back to the well** to migrate from one property management and accounting platform to Yardi Voyager. What is the cost of failure on a software project? With one particular client, who brought us in for their third attempt at implementing Yardi Voyager, we noted the following costs and/or negative outcomes:

- An increased amount of time on their old, inefficient system
- Loss of organizational confidence, including organizational recriminations
- Loss of time to meetings and tactics that were not effective in meeting their goals
- Employee firings, resulting in increased training costs and on-ramp time
- Money spent on software license fees for a product the company was not using
- Money spent on outside consultants with no implementation to show for their consulting investment
- Competitors' pronounced operating advantages from being on a more current software platform

What does it mean to *implement* software *successfully*? The concepts are simple, but their execution is less so. Lupine defines software implementation as when the following outcomes occur *in a certain agreed-to order*:

- The system is set up and configured according to the client's needs and specifications

- The relevant data is converted (and validated) from the source system to Yardi Voyager

- The critical reports are either identified in the system, or created as needed

- The system users are trained on software functionality that is relevant to them

Based on our experience, all of this needs to be done efficiently and effectively *without disrupting the client's existing and current business processes*–meaning the sales and delivery of services/products to their customers.

## SOFTWARE IMPLEMENTATION MYTHS

It is not uncommon during the implementation discovery process to encounter some ill-advised misconceptions about it. Here is a short list:

**Myth No. 1:** You are done when you execute the software contract. Nope–you are just getting started. If you purchase software, but botch the execution of the implementation, then nothing really has happened. Evaluating the right product for your organization is obviously important, but *a skillful execution of your implementation plan is where the financial payoff is*.

**Myth No. 2:** Implementation and training are synonymous. Training is a component of implementation, an important one. But it's only a piece. If you are paying, or plan to pay, for only training, then you will not move more quickly towards your goals. There is more involved than just training.

**Myth No. 3:** Software implementations must take a long time to complete. Not at all true. It's all in the planning. Unfortunately, some companies want to skip this vital step in the interest of moving forward quickly to get a leg up. At Lupine, we have learned that the tortoise always finishes faster than the hare. We have completed migrations for some clients in as little as six weeks.

**Myth No. 4:** Parallel processing between the existing software system and Yardi Voyager must be part of the implementation. Parallel processing is only necessary when you are not certain of the functionality of the system to which you are converting. Yardi's software works; they have a vast and wide client base to prove it. Any sort of pilot testing should occur before the acquisition of the software. Parallel processing is an unnecessary burden on the project and the employees who must do their work in two systems, rather than one, for the duration. Not good.

**Myth No. 5:** Planning can be skipped. Only if you want a guaranteed failure.

**Myth No. 6:** The first step of the implementation is to train all of the users. We get this one a lot. It's a good idea to give the user base an overview of the system, but only for purposes of system setup–in other words, to facilitate system and module configuration. End users should be trained as close to the go-live date as possible. Why? Answer: *Use the information or lose it.*

**Myth No. 7:** You have to be extremely technical in order to convert the data from your system to Yardi Voyager. During implementations, we have taught many a client how to extract source system data and how to import this data into Yardi. If you can use Excel, and you're willing to learn, you can do the import yourself.

**Myth No. 8:** Your implementation project manager must be an *outsider*. The word from the bird is that anybody can be trained how to be an effective project manager. There are pros and cons to hiring an outside professional. The benefits are that you get somebody who is independent, presumably skilled in software implementation, and cannot be fired by internal management. The downside is that they cost money.

## HOW AND WHY IS SOFTWARE DIFFERENT?

The purchase of software is different than any other expenditure your organization will contemplate. It's expensive, it's necessary to run your business, and it must be tailored to your unique set of needs and requirements. If you get it wrong, the cost of failure can be more than the price of the software itself. To add to the burden, your people are almost certainly stretched to the limit, and (this is important) you may not know the proper order and sequencing with regard to running concurrent tasks on a software implementation project.

To complicate matters even further, Voyager is always changing and improving. Which version should be implemented? How do you implement a moving target–a product that is being enhanced constantly? What is a pilot program? How many days will you be without a system? How will your staff be trained? What will the report output look like? How do you get from here to there? These are a lot of fear-based questions. Valid but fear-based questions.

Here are some more items that might be of concern to you:

Is hiring a consultant like opening up your checkbook and writing a blank check? How do you control these costs? What will happen if your customer's data is incorrect? What can be done to insure that this doesn't happen? What will happen if your owner's data is incorrect? What can be done to insure that this is an impossibility?

What if the software does not operate as advertised and demonstrated? What are your remedies? What will happen if all of the key knowledge and information resides with one employee? Can he or she hold you hostage? Is the software company representing your interests or their own during the implementation? How will you know?

## PROTECTING YOUR INTERESTS

Implementing software is not hard. The key, though, is knowing that the required steps must occur in a certain order. Step C must follow Step B which must follow Step A. The failures we have witnessed occur when the organization (in an attempt to save money) skipped from step L to step X to step A, and employed none of the techniques and tactics that should be in every project manager's toolkit.

So, how do YOU protect your interests during the Yardi software implementation process? How do you avoid the potential pitfalls and mistakes?

1. Planning. From time to time, we have clients who want to blow past this step and go "play" with the software. Doing your job using the software is the easy part–it's a good intuitive software product. Determining the use, strategy, and timing is difficult. It takes some measured thinking and some mental rolling-up of the sleeves.

2. Clarity. You need to know what it is you are trying to accomplish, and within what time frame. This clarity needs to exist at an organizational level.

3. Buy-in. Just as with clarity, the buy-in of the upcoming change needs to exist at an organizational level. From time to time, it happens that not everybody agrees with the decision to select a certain software product. That professional disagreement must be put behind you in order to move forward.

4. Finances. Spend money where you need to and not where you don't. You are going to need help as you go through the implementation process. Spend the money on project planning, software module setup planning, data conversion strategies, and training. We see many clients spend a fair amount of coin on creating custom reports that match the reports they are getting from their current system.

5. Economy/Austerity. Don't buy modules that you will not use the first year. Just say "no." Remember that your license starts as soon as you start the software discovery process. Add modules and functionality as they are coming online.

6. Re-think your approach to converting transactional history. Importing and validating transactional history is almost always the biggest time investment for our clients. (It has to be validated–and YOU have to do it.) This step normally has a very low organizational payback, with a high cost in terms of resource utilization. You will see references to this concept over and over in this book.

7. Adopt a phased approach to the implementation. The strategy of having small "victories" stretched out over time is a less risky approach than an "everything goes live at once" strategy.

8. If you are forced to push back a go-live date, do so early in the game. Last-minute deferrals will give rise to a sense of project and organizational failure. Everybody forgets that the date was pushed back if you do it early enough.

9. You must have weekly status meetings that include all relevant parties from within your company, your consultant (if you have one), and the key Yardi personnel working with you on the implementation.

10. In addition to No. 9, a weekly status report must be created that communicates the following:
    - Work completed this week
    - Work assigned and not completed
    - Key decisions made during the week
    - Work assigned for next week

11. You must maintain, monitor, and communicate a project issues list. In the beginning, everything is an issue. You cannot solve problems and issues that have not been inventoried. The failure to keep a list like this is often a large part of why software projects fail.

12. Delegate project tasks when necessary, but never abdicate responsibility (even if you have hired a consultant to help you).

13. Partner with Yardi Systems and treat them as one of your strategic partners. They are good "foxhole" people–ones who will stay with you even if you are going through a difficult time. An adversarial relationship is not good–these projects can be difficult enough without one.

14. Keep a positive attitude and good humor. This IS going to happen for you. Thousands of companies have gone through what you're about to, and most have done so successfully. Keep your eye on the ball and realize that you may have some tough days on the project. That's all they are–tough days. Join the club.

You are embarking on an interesting professional journey. Selecting, designing, and implementing new software is not without its challenges. (In particular, having to remain mindful that all of this is going on while you do your regular job.) My consulting firm, Lupine Partners, has been providing these protective services *with an exclusive focus on the real estate industry* since way back in February 1993 when I founded the company. Since that time, we have successfully provided software consulting solutions to hundreds of clients. While we learn something new with each client, there is very little ground that has not been covered before. At this point, most of our learning is in the nuances and in the margins. In short: we are not rookies.

Okay, let's get started. In the remaining chapters of the book, I am going to discuss everything you ever wanted to know about software implementation. But first, we are going to talk about failure. See you in Chapter 2.

## CHAPTER SUMMARY

1. *When he was a young man, David routinely got his butt chewed out by former Texas Governor (and presidential candidate) John Connally.*

2. *Your main concern at this point should be the cost of failure–of repeatedly having to go back to the well to migrate from one property management and accounting platform to another.*

3. *By definition, software implementation requires that definitive outcomes occur in a certain agreed-upon order.*

4. *During the implementation discovery process, you are likely to encounter some ill-advised misconceptions regarding the process.*

5. *The purchase of software is different than any other expenditure your organization will contemplate.*

6. *Implementing software is not hard. The key is in knowing that the required steps must occur in a certain order.*

# CHAPTER 2

# I'M TALKING ABOUT... FAILURE

It may seem odd that I'm starting this book by talking about project breakdowns and lessons I have learned from them over the years, but I am doing this on purpose. Over the course of my career, this is how I have begun each software implementation project with every one of my clients. I talk to my client about what can go wrong during these difficult undertakings. I believe that knowing about dysfunctional project tendencies ahead of time lowers the probability of these conditions actually occurring during the actual engagement.

While there is no one right way to implement software, there are many wrong ways. And the wrong ways are the focus of this chapter. There is strength in knowing what **not** to do. Several times over the years, I have interrupted meetings to say, "You remember when we started this project and I made you all sit through my ramblings of why projects fail?" "Yes," they say.

"Well, we are doing some of them—and we have to stop," I tell them.

This caution always proves effective because we have the discussion in advance and because we are deliberate in our efforts to curb dysfunctional project behavior. The list I am going to provide is not necessarily a comprehensive catalog of behaviors and actions related to project degeneration. But they are the ones I have seen most often while working with hundreds of clients in matters related to business software.

Here are the major offenders:

- Lack of executive or managerial buy-in
- Projects that take too long
- Organizational resistance to change
- Poor resource allocation
- Negativity
- Lack of a plan
- Rushing
- Cancellation of status meetings
- Forgetting that you're part of an organization
- The perception that "the sky is falling"
- No system for identifying and resolving issues
- Lack of project leadership
- People not doing what they say they are going to do
- No defined project end date
- Lack of delegation

**Lack of executive or managerial buy-in.** The software implementation process must work from the top down. This is a large undertaking in any organization, one that requires hundreds of hours across all functional areas. If executive buy-in is not in place, then team members may not make their best efforts, and the leaders of the company will not respect or support the hours and dedication it requires to get the software implemented.

I put this lesson at the very top of the list because it is the No. 1 reason behind projects "gone wild." Think about it: your organization is going through a major initiative to implement new business software. You are going to invest major time and money in doing so; shouldn't the top managers and executives be on board and engaged?

It is tough to get past poor leadership and do a good job on projects when that sense of buy-in doesn't exist. If the owners or executives of the organization do not care about this project, and have no interest in why you are switching to Yardi Voyager, you might be successful anyway. But there will be times when you are spread thin and the higher-ups need to make the call whether you are working on the implementation project or some other aspect of your regular job. You need these people on board. If they are not, as I said, you can make it but it will be tougher.

**Projects that take too long.** Here's the deal: Most people—and organizations—only have so much time and willingness to work on projects of this magnitude. If an end date is not specified up front or if the project continues to be delayed, the momentum and energy for finishing the project will dry up and go away. Also, company management will lose confidence in the project team. If the project looks like it is going to be a long one, then break it up into bite-size phases and multiple projects so that it is more palatable psychologically to all concerned.

Your organization only has so much bandwidth that you can use to complete this project. You have your regular job, everybody at your company is presumably already busy, and this is all extra. You need to set a definitive end date for this, and you need to manage the project to that end date. If you are taking too long, your company may not have the endurance to make it to the finish line.

**Organizational resistance to change.** Several years ago, I led the implementation effort for a large property management firm in the Southeast. In one debriefing session, I said that it was natural for some procedures to change when implementing a new software solution. A woman who had worked with the company for over thirty years stood up and gave me a blistering commentary about how doing business a certain way had made the company so successful and that going on a new software package shouldn't have to change that.

Her point was well-taken, but she missed the nuance of what I was saying. The essence of a procedure—particularly if it has given you a strategic advantage—doesn't have to change, but the manner of it most probably will. Implementing new software means change—period.

**Poor resource allocation.** This is also known as the failure to realize that people have full-time jobs. Software implementation efforts take time, and project members must have the support of their bosses and superiors. If other work keeps getting piled on a team member, then something will have to give, and it's usually the implementation project. Deadlines will be missed and the project will get pushed back.

This is written for your boss: A strategy of, "Well, we will just work every weekend" is not realistic. Not for twenty-six weeks you won't: you'll probably burn out. This is hard work. As we create custom work plans, we build in some contingency time. If a task usually takes one day, I would probably reserve five days, or even up to two weeks, on the calendar. Part of the art of all this is being honest with yourself and with your organization about how much time you have to devote to this project.

**Negativity.** We've all been around these people. They are incapable of being positive and leave a trail of negativity behind them similar to the cloud of dust that followed "Pig-Pen" in the old "Charlie Brown" comic strip. They are usually profound worker bees—and therefore often end up on the project team. I have seen these people take a fully functioning project team and grind it down to grim death. These projects are hard enough as they are; a healthy sense of humor is a requirement.

I have observed that if you are positive, the negative Nellies of the world will realize that you are not playing by their rules. Then, if you're lucky, they will move on down the road and torture somebody else. These people really are project cancers and should be kept away from the project if at all possible.

**Lack of a plan.** If you don't take the time to plan and include the relevant functional components of your organization in the process, you are most likely doomed for failure. The tasks required to reach your goals must be spelled out in detail.

There have been times where we come in after one or two botched implementations that the client has tried to do themselves. At some point, they have just gone in and "played" with the software. This software is too complicated for you to just experiment with it, and this process doesn't take into account that you have to keep your business running. That flow of information still has to occur; after all, you still have owners and customers, and you have to keep getting those reports out. You need to have a plan.

**Rushing.** You can actually finish faster by going slower. Occasionally I have been unpopular with my clients because I have had to rein them in from things that are not in their own best interests. As I say on my website, "You need to go Step A, Step B, Step C, Step D." Sometimes these steps can happen concurrently, but other times I see clients going to step L then B then Z then J without any coherent reason. *Just go in the order that we give you.* If you botch it, you're either starting over or going back to the middle and you will end up finishing later than you would have. In the old fable, who is faster, the Tortoise or the Hare? Answer: The Tortoise.

**Cancellation of status meetings**. This is more symptomatic than anything else. If you are having your meetings at 10am every Thursday and continue to cancel them because everybody is too busy, that should say something to you. Perhaps you are out of control; perhaps the project hasn't been given the priority it needs. If you cancel two status meetings in a row, hearken back to this page. You have a problem and you need to address it.

**Forgetting that you're part of an organization.** You can be on the bus, off the bus…but there is a bus. This dynamic can also be called "failure to understand who actually owns the company." Twice in recent years, we have had project team members fight the whole notion of the software implementation change *as we were going through the implementation process.* They did not agree with the decision to migrate to a new system, and they decided to fight it kicking and screaming throughout the whole process. It puzzled all of us on the Lupine side, since we did not understand why the executives of the organization put up with such nonsense. At Lupine, everybody is encouraged to give me their thoughts, but once I make a decision they are all expected to be, well, on the bus. Otherwise, they can catch a cab. If you, dear reader, see this behavior in somebody on your team, give them one chance to get with the program. Otherwise, boogie on down the road without them. Do not negotiate with hostage-takers!

**The perception that "the sky is falling," or a lack of belief in the project plan**. If you have a plan and it's a good plan, it's going to work. I understand this may be stressful for you, but you have to believe in the project plan and not flip your wig the first time something goes wrong. It is inevitable that you will be thrown some curveballs throughout all of this. No matter how well you plan these things, it could be as simple as a key person on the team falling ill and being out for a month. It's a curveball. We didn't plan for it, and couldn't have planned for it, and we will have to make accommodations. "It's fine, we will adjust," is how you need to look at it. Be calm.

**No system for identifying and resolving issues.** Let's file this under a lack of project discipline. Here is what probably happens on every project: People in your organization may get together, whether out at lunch or just hanging out, when they start talking about things. Just start filling in the blanks. If you are asking a question, then it's an issue. Talking about an issue without writing it down and capturing it so that it can be resolved does nothing. It's a useless activity. You have to be disciplined enough to write all of these things down. Not every issue is going to be a software issue. In fact, I would say that most are not going to be software issues. Instead, they may be organizational issues involving staffing or time frames. It might be an issue reagrding training or one about extracting data from your source system. But whether it's a question or something that's bothering you, write it down. There are going to be templates for you to look at, but there is no magic here. We just use Excel. We prioritize the issues as "high," "medium," or "low." When they are resolved, we note when and how they were handled. Alongside the project plan and the status report, it is our way, of keeping control of the project.

**Lack of project leadership.** Every project needs a steward. A common error is the failure to designate a point person for the project. It doesn't matter whether the person is hired externally or appointed internally. What is important is that somebody is recognized as the lead on the project.

There are a couple of different ways to obtain the leadership you need. You can locate it or cultivate it internally or hire it out by engaging a professional consultant. But you need a point person, and you should realize that it will be a part-time, minimum of twenty hours per week gig. It is entirely likely that the project manager's normal workload may need to be reduced during the project. One of the dangers with internal project managers is that they are given tons of responsibility with very little authority. *No bueño.*

If you are the project manager of software implementation effort and are reading this book, then you will receive many tips on how to effectively lead your organization through the process. It's at the core of this book's message. Read on!

**People not doing what they say they are going to do.** Or, even worse, people not telling other people that they're not going to do what they said they were going to do. When team members do not perform, it causes two things to happen, both of them bad. One, it hurts morale. You broke the implicit covenant that you made with the other team members: that the project matters to everyone and that everyone will pull together to make it work. Two, it has the possibility of pushing back the implementation go-live date or compressing the amount of time in which other tasks need to be completed.

Life does intervene during these projects. All of us have other obligations, both personal and business. If you are not going to be able to meet a previously agreed-to deadline, then let your project manager know. The project manager will then be responsible for altering the plan, if necessary, and making the required communications to affected parties.

Otherwise, this is what happens: You are in the middle of the implementation and you have handed out tasks in your weekly status report. Say Jim has three items due, and at the end of the week, he hasn't completed them. Even worse, Jim didn't tell you he wasn't going to do them. You can't do the next step until Jim's three tasks are done. His failure to complete them doesn't doom the project, but at some point you will have to postpone that go-live date because Jim didn't do what he said he was going to do.

**Lack of delegation**. Sometimes, it happens that you get a team member who gets immense professional satisfaction from working on the implementation project or is just the type of person who wants to complete all of the tasks himself or herself. That's a good thing, but it creates a problem when you go live: if one person has done everything, then that person can only do one thing at a time. Sometimes at go-live, you may have three or four activities going on at one time. Be aware of this: it is both good and bad. If one person is doing everything you will probably have a slower go-live time frame than if you have two or three people sharing the duties.

**No defined project end date.** If the end date of the project is up in the air, then there will be no drive or immediacy to any of the steps required for the implementation and the product acquisition. The project team members will run out of steam. All of us need goals to operate at a high level, and project teams are no different. A few times, I have been hired at the midpoint of project to help get the client on course. Not having a definitive project end date is invariably a factor when a software implementation project suffers.

From this point forward in the book, (most) everything presented will be positive. I have presented the above lessons many times in my career— in hotel ballrooms, in auditoriums, and many times in a client's small conference room. Recognizing unsuccessful and unprofitable software implementation behaviors is important. You need to self-monitor. If you don't, I will...

In the next chapter, I will discuss the software implementation *planning* process and why it is so crucial to the overall success of the project. Catch you on the flip side.

## *CHAPTER SUMMARY*

*1. Knowing about dysfunctional project tendencies ahead of time lowers the probability that these conditions will actually arise during the actual implementation process.*

*2. The major reasons why software projects fail, or are more difficult than they need to be, are:*

- *Lack of executive or managerial buy-in*

- *Projects that take too long*

- *Organizational resistance to change*

- *Poor resource allocation*

- *Negativity*

- *Lack of a plan*

- *Rushing*

- *Cancellation of status meetings*

- *Forgetting that you're part of an organization*

- *The perception that "the sky is falling"*

- *No system for identifying and resolving issues*

- *Lack of project leadership*

- *People not doing what they say they are going to do*

- *No defined project end date*

- *Lack of delegation*

*3. Warning: If you exhibit any of the above behaviors during the software implementation, David may bust your chops.*

# CHAPTER 3

# WHAT DOES IT MEAN (TO BE ON THE TEAM)?

Remember the Fram oil filter commercials from the 1970s with the tag line, "You can pay me now, or you can pay me later"? That same logic applies to software implementation. If you don't have the proper project infrastructure, then you may hit a point where you unnecessarily elongate the implementation timeline ... thus, paying later.

All implementation projects have a project team. Most people think erroneously that this term refers just to the people in the client organization who are responsible for building the new system, but the client team is just a subset of a larger team which will include Lupine Partners' representatives and key personnel from Yardi Systems. While there is always a project team, there is also sometimes a steering committee on larger implementation efforts. The steering committee should be comprised of the executives or top managers of the organization. They will not be involved in the hands-on implementation process, but they will be considered the owners of the project. As a result, they must do the following:

- Designate the software implementation team.

- Approve the budget for the project team

- Resolve deadlocks for the project team (if they arise during any part of the process)

- Allocate time for their subordinate(s) to work on the project
- Receive periodic status reports from the project manager

What they must NOT do:

- Meddle in the implementation team processes

You may hear the terms "steering committee" and "project sponsor" used interchangeably. Sometimes they are the same people or group of people, but not necessarily. The project sponsor is just that: the person or group who initiates and champions the project. The sponsor may then establish a steering committee to oversee the project team.

Two engagements I worked on really stand out with regard to my experience with a project sponsor. In both instances, the president of the company was present at the kickoff meeting and gave a speech to the effect of, "This project is very important to me and I feel that a successful effort will give us a competitive advantage in the marketplace. David has my blessing to lead this project. Give him your full attention and cooperation." How sweet is that? These projects both went really well. Everybody fell into line and worked hard. We finished on time and on budget—in large part because those CEOs were smart and acted as good advocates for the project.

## THERE'S NO 'I' IN TEAM

In any organization, one person cannot and should not "do" the entire software implementation. When selecting the project team, the steering committee should consider the following:

- All functional areas of the organization should have representation on the team.
- Existing workloads. For example, if the ideal candidate from Operations is going to be traveling periodically over the next six months, maybe he isn't actually the best candidate.

And the following questions should be answered:

- Who is the best person for the role of project manager?

- Do team members need to have experience with software?

- Should the position be contracted out?

- How large should the team be? What is the optimal size for a software implementation team?

- From what level(s) of the organizational chart should the team members be chosen?

- What is the motivation level of possible team members? Do they want to serve? Are they TOO eager to serve?

- How long has the potential team member been a member of the organization?

- What will the responsibilities of a project team member be?

- What will the workload be?

Here are my thoughts on these questions:

**Who is the best person for the role of project manager?** Answer: It's usually obvious. There always seems to be somebody who has the right temperament and experience to navigate the myriad of issues and personalities. Generally, a good project manager is:

- Calm

- Organized

- Good with difficult people and situations

- Slightly reluctant to take the role

- Inclined to run tight meetings

- Both tough and tender-hearted

- Respected by the majority of the people in the organization needed to get the job done

This role should not necessarily be assigned by job function; rather, base the decision on the characteristics I list above. The IT manager may not be the best person to serve as the foreman of this effort. An IT manager or controller who runs a lousy meeting and is obnoxious will make this effort more difficult than it needs to be. Instead, for example, use the senior, savvy operations manager who may have no software experience whatsoever.

Also, the project manager should not play a "rah-rah" role. That is reserved for the project sponsor. The project manager is there to execute and facilitate the project, not to cheerlead the team.

**Do team members need to have experience in software?** Answer: No. They need to have experience in your organization. They need to feel the organizational pain. They need to know what doesn't work as well as it should due to system deficiencies.

**Should the position be contracted out?** Answer: Maybe.

- Consider hiring a consultant to serve as your project manager if: there are no obvious candidates within your organization

- Your organization is tight on resources and you don't want to overload staff personnel

- You want an independent third party to be your project leader

Don't discount the value of having somebody outside your company run the engagement. On numerous occasions over the past twenty years, clients have said to me, "Would you go talk to Person X? He will listen to you." Or, "You and I say the same thing, but because we're paying you the high hourly rate, they listen to you." Another benefit to having outsiders come in is that we don't have to care about company politics. We do our job, get paid, and go home. Sometimes an internal project manager is stuck running the project in a manner that doesn't make sense because of the realities of his particular organization (in other words, because he wants to keep his job).

**How large should the team be? What is the optimal size for a software implementation team?** My preference is five to seven people. That number usually covers all the functional areas of most organizations and allows for the meetings in which everybody can be heard but that do not go on too long. As the number of participants grows, so does the length of time to conduct a meeting, particularly if the meeting is poorly run.

I was hired to manage a software evaluation project for a large East Coast property management firm a few years back. The leaders asked for volunteers to serve on the implementation team, and they got eighteen responses, fifteen of which were from the same functional area. All eighteen people ended up serving on the selection team. It was big-time overkill, and it made conducting the planning meetings and the subsequent vendor demonstrations very difficult. There were reasons why my client did it—over my objections–but the ultimate selection was not "better" because the selection team was too large.

**From what level(s) of the organizational chart should the team members be chosen?** They should be neither frontline staffers nor executives, but rather employees from the middle. We want people who know the organization and are used to working hard. Everybody tends to know who these folks are; they are the ones carrying the load for everyone else! Ambitious types who want to serve just so they can say they did should be avoided. These guys will not work enough, which will breed anger in resentment in the other team members as they have to pick up the slack of a non-contributing team member. Having said this, I have had people on all rungs of the org chart be successful on a team. They have to want to work, though; that's the key characteristic.

The internal team members should consider their roles to be like those of senators. While they may be representing Accounting, Information Technology, Operations, or whatever their individual department is, any decisions *should be made with the entire organization in mind*. This means putting the goals of the project ahead of your departmental goals. At times the goals may not be in concert. On many occasions I've seen client team members make a decision that means a lot more work for them and their sub-team members, but is in the best interest of the organization. It always warms this old man's heart…

**How long has the potential team member been a member of the organization?** I recommend avoiding somebody who just joined the company. All he or she is going to do is bring biases from his or her previous employer. You need somebody with some institutional knowledge.

**What will the responsibilities of a project team member be?** Here is a summary for your consideration:

- Make time to attend meetings

- Keep an open mind

- Do what's best for the company, not just his or her department

- Get his or her regular work done too

- Work hard

- Enjoy the professional challenges that arise

I have worked with some excellent professionals over the years who have told me years later that serving on the implementation team was one of the best, most challenging, experiences of their professional careers.

**What will the workload be?** Usually, it will be quite heavy at the beginning, as you go through the planning process, and at the end, when you are going through the vendor demonstrations. Estimate eight hours per week for each team member during the duration of the process. Triple that for the project manager. Scheduling team members' time is one of the more challenging aspects of the project manager's job. The further out you can earmark weekly time for the team members' commitment to the project, the better. If you can book meetings well in advance, then you get on team members' calendars.

Having the *right* people on an implementation team is critical if you want to have a low-stress, on-time migration effort. (Go back to "Why Projects Fail" and re-read the sections on Negativity and Forgetting You're Part of a Company.) Having the wrong people along for the ride can be hazardous to your health!

In the next chapter, we will burn rubber and start "making plans."

## CHAPTER SUMMARY

1. *If you don't have the proper project infrastructure, then you may hit a point at which you unnecessarily elongate the implementation timeline.*

2. *All implementation projects have a project team.*

3. *The client team is a subset of the larger team, which will include Lupine Partners representatives and key personnel from Yardi Systems.*

4. *The project sponsor is the person or group of people who initiate and champion the project.*

5. *The project manager is there to execute and facilitate the project, not to be a cheerleader.*

6. *Project team members should be neither frontline staffers nor executives, but from the middle of the organization.*

7. *The internal team members should conduct themselves like senators do. While they do represent their individual department, decisions should be made with the entire organization in mind.*

8. *Project team members should estimate an average time commitment of eight hours per week during the duration of the process. (Triple that number for the project manager.)*

# CHAPTER 4

# WHAT AND WHEN

It's time to get this software implementation party started. We always do this by facilitating two meetings: the discovery meeting, which I'll address here, and the project kickoff meeting, which will be discussed in the next chapter. The discovery meeting is the initial strategy session for the entire software migration and implementation process. We typically hold this meeting on a Monday, and its purpose is to determine the "what" (or "which") and "when" of the implementation: *Which* modules, *which* interfaces, and *which* functionalities, and *when* should the migrations take place. The outcomes of this session should be:

- Organizational consensus on the scope of the project
- Eagerness to move forward
- Relief at getting the process started

These meetings are typically held in the client's main conference room and should involve the participation of the entire management team–or, at the very minimum, the project team. The goal of the session is to get enough information about the "what" and "when" of the project to craft a custom solution and work plan for it. The goal is NOT to solve problems or to determine *how* the implementation is going to occur. That will happen later. I have had clients who have elected to skip this discovery session. In every such instance, there were problems during the implementation because there was not up-front organizational agreement about "what" was being implemented. The results: finger-pointing, wasted time, and recriminations. If you skip this session, you are tripling your risk. These discovery sessions do not need to go on for days or weeks; you may be able to accomplish all this in a ten-minute meeting.

## YOU CANNOT IMPLEMENT A MOVING TARGET

The first reason you have a discovery session is to determine the scope of the project. The scope is the "What." **What** is it that we are going to be implementing? You have bought Yardi Voyager for a reason. I have never had an instance in which I have had complete and immediate agreement among the members of the client project team at a discovery session. For instance, person A thinks you are implementing General Ledger and Accounts Payable. Person B thinks you're implementing General Ledger, Accounts Payable, and Residential Management, and a third person thinks you're implementing Job Cost. Often a great deal of time passed since you first decided to buy Yardi Voyager and signed the software contract. It is extremely possible and highly probable that there will not be a meeting of the minds as to the project scope. So one of the hoped-for outcomes of this discovery session is to reach organizational consensus as to the "what" of the project.

You cannot implement a moving target. Right at the beginning of the project, we want to draw a box around what is going to be implemented. I have led projects in which halfway through some new functionality gets added to the project scope. Sometimes it can be handled as it arises, but often we are forced to backtrack because the addition has an impact on other tasks that are dependent on the affected task. One of our goals in this session is to get agreement on the scope—what the project "is"—so that we can implement something that is standing still.

A mistake that I see from time to time during these sessions is people wanting to get too detailed too quickly. All we need to do is at this point is establish a scope and gather information so that we can create the summary and detailed work plan. All of the other details will follow further down the line. So what is a project scope? The project scope is a document that describes what the project *is*. The project scope document does not concern itself with *how* the implementation will be tactically executed or the timeline for it, but rather with *what* is actually being implemented. Here is a real-life scope document we created for a client a few years ago (the names have been changed):

## PROJECT SCOPE

Conversion and setup of Madison Partners' accounting and property management information from MRI to Yardi Systems' Voyager 6.0 platform. The following modules will be implemented: General Ledger, Accounts Payable, Property Management (Commercial, Residential, Condo/Co-Op/HOA, and Affordable), Job Cost, Maintenance, Mobile Work Orders, Executive Dashboard, and Budgeting & Forecasting. Included in the scope of this project is the:

1. Conversion of the 12/31/10 balance sheet (closed to retained earnings) from MRI to Yardi by entity, by chart of account number, by accounting book AND the conversion of monthly General Ledger detail transactions by entity, by chart of account number, by accounting book for fiscal 2011 up to the go-live date.

2. Conversion of General Ledger budgets by entity, by chart of account number, by accounting book for 2010, 2011, and 2012.

3. Conversion of all vendors who received payment from Madison Partners from 1/1/2010 to the present.

4. Conversion of open payables by property/entity, by vendor as of the go-live date.

5. Conversion of all commercial current, future and past tenants (who either have a balance owed to Madison Partners or were billed CAM estimates in 2011 from MRI to Yardi). Included in the conversion will be tenant balances by charge code (including prepayments), recurring tenant charges, and security deposit balances.

6. Conversion of all HOA and Residential current residents, and past residents with a current balance owed to Madison. Included in the conversion will be the owner/resident balances by charge code (including prepayments), recurring charges, and security deposit balances.

7. Implementation of the cost recovery functionality in Yardi for purposes of using the system to calculate and bill the tenant's share of operating expenses as of 12/31/12.

8. Conversion of Condo and Rental history for 2010 and 2011 to be accessed via the Resident screen.

9. Implementation of the following Interfaces:

   - Transfer of funds – AP
   - Invoice upload – AP
   - Cleared checks – AP
   - Positive Pay – AP
   - Prenote – AP
   - Receipts import – Commercial
   - FICS import – Commercial
   - Export to eStatements with link – Condo
   - SharePoint tenant statements – Condo
   - ADP import – General Ledger
   - Billing to sites for payroll reimbursement – General Ledger
   - FAS journal entry – General Ledger
   - Recurring journal entry import template – General Ledger
   - Development manager – Job Cost
   - Banking lockbox file – Rental
   - Direct debits – Rental
   - Property Solutions portal – Rental
   - Online application – Rental

- Level 1 call center – Rental
- LexisNexis credit screening – Rental
- Cash receipt import – Rental
- Utility reimbursement import - Rental

10. Implementation of Yardi's system notification functionality.

11. Conversion of all historical work order records from MRI to Yardi.

12. Conversion of all active jobs into the Job Cost module as of the go-live date.

13. Conversion, if possible, of historical Rental guest traffic from MRI to Yardi.

Specifically excluded from the scope of this project are the following items:

1. Implementation of Yardi's Portfolio VMF Module
2. Implementation of Yardi's Residential Portal

––––––––––

You can see this is a short two-page document, not that intimidating in length. Not only does it outline the scope, it specifically excludes a few items that the company couldn't agree upon, since I wanted to make sure everybody knew what wasn't going to happen. This document put a box around the project for this particular client.

You can see this is a very specific, detailed statement: "These are the modules…here are some of the nuances…this is what is excluded." After our discovery session we created this document and came back to the client and said, "This is what we heard, please confirm this," and they did. When I get to this point, it's not uncommon to have a requested tweak to the scope document. What is most important is that everybody in the organization or on the project team says, "Yes, this is it." We will not move forward until everybody is in agreement on it.

## DON'T ASK ME NO QUESTIONS,
## AND I WON'T TELL YOU NO LIES

How should these sessions be conducted? What we do is get the major players, which tends to cover all functional areas of the organization, in one room. As I mentioned before, the truth is that once you get everybody together and start asking questions, you are not going to get an easy agreement. You just sit there, listen, watch, and facilitate. The answer and the scope will present itself. It just happens. The conversations go something like this: "Well, I thought we were doing 1099s." "No, we are not doing 1099s because…" "I thought we were going to do all the properties?" "No, we're going to do all the properties and we're going to do corporate entities as well." And it's beautiful to watch because we're going through this right from the beginning. It's the gestation of a software project. Doing this now ensures that you won't have to go through it again halfway through the project, or even have to start over.

Remember what our goal is: to determine a scope and gather some additional information so that we can craft an implementation plan. Everybody else's agenda may be a little different, but ours is to run this project.

## DISCOVERY TOPIC POINTS

Now that we have everybody in the room, we ask our first question: "What did you buy?" Let's confirm what it is that you purchased and, more importantly, which of the purchased modules you want implemented immediately. Unfortunately, some of my clients buy modules and wait two or three years to implement them, while paying a license fee the whole time. Be that as it may, get that confirmation—because once again, it is very possible that people on the team will have thought you licensed a certain module and you actually didn't.

Next item: When is it that you want to convert? Everybody has some idea about this. Just because they have a thought about it doesn't necessarily mean that they're all going to be right. What are people's ideas? Do we want to do this all at once? Do we want to do General Ledger first and other modules later? Do we have some other business deadline coming up that is going to affect this? To me, the answer to the timing question is *the most important one*—the one I'm dying to hear when I come in. I've worked on major conversions where the client wants to get it all done in six weeks even though it's a nine-month job, and I have to change those expectations.

You will definitely find that people have different ideas, and the goal here is to facilitate that conversation and see where that goes. I've written work plans that were not based on the client's initial desires, because the client's plan was not in their best interest. Another question that we want to ask is: Are you interested in project phasing? Almost all of the jobs that I run end up being implemented in phases. An example would be: Accounts Payable and General Ledger will be implemented first, and then Property Management shortly thereafter. Accounts Payable and General Ledger tend to be easier to implement, and while that implementation is underway, we can orchestrate the roll-out of the properties (particularly if there are a lot of them).

This phased approach keeps different people in the organization on the project team busy but also gives you an early success. You get to say, "Yes, we are already up on Yardi now—granted, it's just General Ledger and Accounts Payable for now, but we're up and running." Another reason to consider phasing is there may be some pressing business need. You just bought a portfolio and you want to get them up on Yardi, so those five new properties go up first and the rest will follow later. Or let's say it's December, and at a minimum you want to go live on Accounts Payable by January 1st so you can capture all of the information for 1099s and you don't have to do a conversion mid-year. My point here is that you don't have to do all the modules at once.

## GENERAL LEDGER

The first area for consideration is the chart of accounts and the question that needs to be asked is: Do you need to change your chart of accounts now that you are going onto Yardi Voyager? Yardi doesn't require you to change your chart of accounts, but I will say that more than 50% of the time the organization decides to because they are unhappy with their previous chart of accounts.

To repeat, you don't have to change your chart of accounts, but if you decide to do so, it has ramifications when you begin converting data. It is an important tactical activity to consider. Don't let this conversation disintegrate into *what* the chart of accounts is going to be. There will be a whole series of meetings about that question. Now, you just need to determine: Are you, or aren't you going to change the chart of accounts? Alongside this, but separate from it, is the conversion of General Ledger history discussion. Basically, we are asking: Are you bringing a balance forward entry by General Ledger account by property, or are you bringing over the history? If the answer is the latter, how much history, and will it be in detail or summary? So the easiest answer is that we will enter a balance forward as of the go-live date and we will refer to the old system for the General Ledger history. The disadvantage is that you will not be able to run comparative financial statements off of Yardi Voyager.

Even though this is your party and you can cry if you want to, please understand that I have seen some projects fail because of an immature decision around the conversion of General Ledger history, and I've seen many of them delayed. Here is how it happens: As they sit in this discovery meeting, people will say "I want all General Ledger history," or "I want five years of General Ledger history," either of which is fine. But if you convert it, you have to balance it. What that means is you will be running reports out of your old system and out of Yardi Voyager to make sure that the data conversion effort was successful. It is even worse if you changed your chart of accounts, because you will be running reports out of your old system in one chart of accounts and Yardi Voyager in another, and often the accounts are collapsed or consolidated. It can happen that the conversion of General Ledger history never catches up to the project, and accountants end up working overtime and on the weekends for something that shouldn't have been all that crucial in the first place. *This is a discussion you want to have.*

## ACCOUNTS PAYABLE

A major consideration is whether we are going to convert year-to-date vendor payments. Let's say we convert as of April 1st. The consideration is whether we are going to convert the January 1st to March 31st payments that have already been made to vendors. There really is no right or wrong answer here, but it does have 1099 ramifications. So what we need to consider is whether we are going to produce 1099s for the first three months of the year using the old system and the following nine months from Yardi Voyager, or instead convert the payments for the first three months from the old system to the new system and produce the year-end 1099s out of Yardi.

Will you have a positive pay interface to your bank? Positive pay is the creation of a check register file that is sent to your bank. They will match that file against checks that clear for security measures, to make sure somebody is not double-using a check in a series. Also, we need to consider whether we are going to use inter-company accounting. If your organization has an environment where you have a payor who pays on behalf of others and you want to create a due to/due from relationship from that payor to the other entities and or properties, this is something that needs to be considered and be included in the custom work plan.

Also, we will ask you whether your organization is going to be using purchase orders as a means for requisitioning materials and services in front of the Accounts Payable invoices. This will also have an effect on the custom work plan.

## RESIDENTIAL MANAGEMENT

You will be asked whether you are going to renumber units, whether we are going to be changing the charge codes (which will affect our conversion at go-live), and whether there is a change in the unit type methodology.

Amenity-based pricing is a residential topic with which you can price individual items or amenities in a residential unit like fireplace, view, balcony, second floor, microwave. You have your base core rent and then you can add on these amenities. Yardi can do that, but it affects how we write the implementation work plan. Once again inter-company accounting questions will be asked. Is there a property or entity that is receiving moneys for others? We also ask questions about the possible interface between external residential sub-systems and Yardi Voyager.

Examples of external best-of-breed systems are utility reimbursement, credit checking, and Internet listing services.

## THOUGHTS ON PROPERTY ROLL-OUT STRATEGIES AND PILOT PROGRAMS

We have led implementations with up to two hundred properties, and with that many properties you can't do the property roll-out all at once—from both a data conversion and validation standpoint and from a training perspective (training the end-users). It's too big of a bite. What I like to do here is simply ask if people have any thoughts on the subject. Maybe there are properties that are going to be excluded. This isn't the session for the mix to be ultimately decided, but I like to hear the discussion because it may affect how we write the work plan and how we create the entire migration approach.

There should be a discussion of whether you want a pilot program. The purpose of pilot programs is to test to see if the system is going to work as advertised. Well, you've chosen Yardi, which has an enormous install base—the software works. Pilot programs generally don't go well because there isn't enough time to do conduct a proper program. Now, you may want to pilot maybe certain aspects of the program, if there are some nuances to your organization. But as your consultant, I would advise you that it is still the wrong time and is the pilot is something you should have done prior to your actual purchase of the software. It is an evaluation technique, not an implementation technique.

## JOB COST

We ask about what kind of construction jobs are going to be implemented—tenant finish-out, regular construction, rehab, etc. We also ask whether your organization has a chart of categories, and if so, is that chart going to change because of the Yardi conversion? Will we be creating draws within Yardi, and is there is a desire or expectation that we will convert the data on existing jobs that are in construction? It is a difficult thing to convert in-process jobs because of subcontract and retention nuances. Most companies decide not to do it.

## COMMERCIAL MANAGEMENT

In the commercial area, all of the residential points apply, and you also want to find out your organization's expectations with regard to recoveries. I've never done a recoveries implementation in the first phase. There are two aspects to recoveries: You have the expense pools on the General Ledger side, and then the actual set-up on the commercial management side. It is a Phase Two endeavor, and it is always difficult to do it if you are coming on mid-year. We will also ask you questions about workflows, prospect/deal management, straight-lining of rent, CPI, tenant ACH payments, and square-foot history.

## SUMMING UP

When I am done with my questions, I restate my perceptions of the discussions. An example: "Here is what I heard you say…you are going to be implementing Accounts Payable, General Ledger, Property Management, and Investment Management with a proposed go-live date of June 30th. You will be totally redoing your chart of accounts. You will be bringing in two years of General Ledger history. You will be bringing in year-to-date vendor payments because you want to do all of the 1099s out of Yardi. You are going to be doing the equity roll-up in investment management. You will not be reconverting tenant history, but you will be renumbering our units and you want to do ten properties in June and ten properties in July."

I say "Is that it, generally?" Then I look around the room and say, "I am going to formalize this for all of you but for right now, we are going to leave and start creating our work plan. Is that it? Did we get it? Is everybody in agreement? Speak now or forever hold your peace." All of the heads should be nodding up and down. This may be the most important thing I can do during this entire implementation. I rarely get one hundred percent consensus during that restatement. There is generally at least one item about which people just see things differently.

In the next chapter, I'm going to talk about what we do on the Tuesday, Wednesday, and Thursday following the Monday discovery meeting.

# *CHAPTER SUMMARY*

1. *The discovery meeting is the initial meeting for the entire software migration and implementation process. The purpose of the meeting is to determine the what and when of the implementation.*

2. *You cannot implement a moving target.*

3. *The project scope is a document that describes what the project is. The project scope document does not concern itself with how the implementation will be tactically executed or with go-live dates, but rather with what is being implemented.*

4. *You want to decide the project scope at the beginning of the process.*

5. *Not everybody will agree on the scope of the project at the beginning of the discovery meeting. Respectful professional conversations will carry the day and culminate in organizational consensus.*

# CHAPTER 5

# OPENING KICKOFF

The discovery meeting is over. You get to go back to your desk and catch up on all of the work you missed while attending the Monday session. We at Lupine, however, are required to be back at your office in two-and-a-half days to present the following project materials to you on Thursday:

- Scope document (Chapter 4)
- Why Projects Fail (Chapter 2)
- Summary work plan
- Detailed work plan
- Issues list
- Short-term tasks going forward

The kickoff meeting is the official start of the software migration and implementation project. It is one of the most important days of the project: Enthusiasm is high, anxieties and fears are reduced, and the map of how you are going to get from here (status quo) to there (migration to Yardi Voyager) is communicated and (hopefully) accepted.

Historically, attendance at these meetings has ranged from anywhere from two people to 500. I have been hooked up with a microphone and presented the materials in an auditorium…and I have casual conversations with a handful of people in a conference room. This is an opportunity, however, to hear how this project is going to go, straight from the horse's mouth. In my opinion, the greater the number of people in the organization who hear the plan the better. The software migration effort will affect everybody in some way and at some point as we move forward together. A few companies elect to have the project sponsor stand up and say a few ceremonial words before we begin presenting the project documents. These words are usually powerful and profound. This gesture gets people in the organization aligned and focused. I always ask at the Monday discovery session if somebody is willing or desires to give the opening words. Rarely do I get any takers. It's unfortunate because it is a missed leadership opportunity.

## VISUALIZE WHIRLED PEAS…

Your role in this meeting is to sit, listen, visualize, and to **think**…and to ask questions as they come to you. When the session is over, you should have a good sense of the strategy and approach that has been crafted for you. At Lupine Partners, we begin these sessions by discussing failure, holding a 20-minute discussion of things we have seen other clients do throughout the years that have made projects more difficult than they needed to be (or in a few cases, where the entire effort had to be unplugged due to unprofessional attitudes and actions). (See Chapter 2.) After that, everything else is positive.

Here is a **sample kickoff agenda** for your consideration:

1. Introduction and Purpose
2. Why Software Projects Fail…or Become More Difficult than They Need to Be
3. Confirmation of Project Scope
4. Presentation of Implementation Strategy
   a. Five Challenges:
      - 23 interfaces to build/test
      - Large number of people to be trained

- Trainer skill set and trainer not identified
- Many different modules
- Every department wants to go first

b. Summary Timeline and Presentation of Suggested Implementation Approach

c. Detailed Work Plan

5. Tasks Going Forward

a. Send project documents to Yardi account manager - Lupine

b. Hold coordination meeting with Yardi account manager
   - Lupine

c. Set up Yardi environment (load packages, service packs, and plug-ins) – Client

d. Load sample database with module design setups
   – Lupine/Client

e. Attend upcoming module design and configuration sessions
   - Client
   - What to bring
   - Lupine representation

f. Begin GL account mapping between MRI and Yardi - Client

g. Complete chart of accounts after November 14th meeting
   - Client

h. Make general assignments for data conversion and interface projects - Client

i. Bring lease expiration reports to 11/30 data conversion meeting for purposes of discussing "issue" tenants - Client

j. Determine skill set for Yardi trainer and drop-dead date for hiring - Client

## STARTING THE MEETING

First, we will re-state the project scope as discussed in the discovery session. The kickoff meeting will not move forward until everybody is in agreement as to the scope of the project. As you learned in the previous chapter, *it is impossible to implement a moving target*. Once that is done (and once we have discussed project failure!), your custom solution approach will be presented. This is typically done with two different documents: a summary timeline, and a detailed implementation work plan. We do not move forward until everybody understands the plan.

The summary work plan is a one-to-three page, landscape-oriented, summary flow of major tasks and milestones. It is meant to tell the story of how the project is going to be approached. It is the document we create to communicate how the project will be completed. The plan usually encompasses a 4-6 month time span. Most of the kickoff meeting is dedicated to going over the summary work plan. We use the detailed project plan as the tool with which we actually manage the project. These two documents are equivalent. They both encompass the same project approach and implementation time frames–they just serve different functions in the project planning process. The detailed work plan will normally list between 250-400 individual work tasks and will be anywhere from 10 to 15 pages in length.

## IT'S A SECURITY BLANKET

When I sit down to create the summary plan, I know that I have a puzzle to solve. How do I meet the client's desire to have certain modules implemented by a certain date with only a certain number of client personnel available to work on the project? Sometimes the answer comes to me immediately, but sometimes not. I have, on occasion, gone back to my client and said, before the kickoff meeting, that they cannot have what they want. It's not doable. A nice by-product of the process of solving this puzzle is that, when we present the detailed work plan to you, we know the project backwards and forwards. We can stand up in front of our client without looking at any notes and say "This is how we are going to do it. These are the time frames…"

The detailed work plan inspires confidence. It shows others that we know how we are going to get this done. People see us carrying it with us, taking it into meetings, and constantly marking it up. The work plan lowers everybody's anxieties (yours and ours). Occasionally I have to get up in front of people and speak. I always have some sort of cheat sheet in my back pocket, just in case my brain goes blank in front of all those people. This is our cheat sheet. It is always there for me to refer to, and serves as my security blanket. The detailed work plan is a throw-away deliverable. It is a means, not an end. It is just a tool. I've seen other consultants treat it as a holy grail, and it's not. The goal is not to have a terrific work plan; the goal is to be converted to Yardi Voyager.

## IN THE BEGINNING, EVERYTHING IS AN ISSUE

A presentation of the issues compiled to date will follow. During the discovery meeting, and while we were creating your project documents, certain issues will have arisen. This is normal. At this point in the project, it is more important to capture the issues than to resolve them. The resolution will come later. In fact, it will happen over the course of the implementation—repeatedly.

I have carefully trained all of the consultants at Lupine in the art of listening for issues. Within two minutes after the discovery meeting begins, issues will start to arise. An issue is a project-related (not necessarily just software-related) question that needs to either be answered or considered. And not all issues are equal. They must be prioritized from "critical" to "low" importance. For example, consider these three issues:

- Is the training schedule changing because of redo of property mix?
- Does third party management require separate chart of accounts?
- Where do we set up the months to start concessions?

The first two issues are critical, in that the decisions surrounding them will affect multiple tasks going forward in the project. The third issue is just a task showing this particular user how to do something. But all three are valid, viable issues. The first two got our attention and required immediate attention. You can't solve issues unless you inventory them. This compilation effort goes on for the duration of the project, but most of the issues are uncovered by the time the module design and configuration meetings (Chapter 6) are completed. The resolution of the issues comes as a matter of course over the life of the project. Sometimes, though, we will call special issue resolution meetings between Lupine and Yardi to work through some of the thornier problems.

Here are some other examples of real-life issues:

- Can multiple bank accounts be selected when issuing checks?
- When selecting invoices, what is the best way to ensure the full invoice is selected when the invoice has been allocated to multiple properties?
- When selecting invoices, what is the best way to ensure that there is enough money to cover all invoices selected?
- Can tenant statements be run for a specific charge code?
- What shows up in the tenant ledger when a work order is billed back to the tenant?
- How is the tenant data going to be extracted from MAS?
- How many characters will be used for the new chart of accounts?
- What will the required segments be for the departmental accounting?
- How will the existing leases be brought over if the tenant has more than one suite?
- Can property assets, such as HVAC units, be imported into Yardi as "unit assets," rather than manually entered?
- What is the best way to track expenses when there are two buildings, and some tenants pay on the total electric for both buildings, while others pay on electric for only the building they are in?
- Are there enough resources at the client company to complete all the mapping and validation requirements?
- Is the module that will allow credit card payments an add-on module or part of a core module?
- Can one scanner be used to scan checks for accounts at multiple banks?

The kickoff meeting will end with us discussing "going-forward" tasks. These are key short-term tasks that are going to need to be addressed as soon as the kickoff meeting ends. Know that most of these tasks have nothing to do with the software per se. They are project tasks that are relevant to the go-live date that was communicated in the kickoff session. The software sale may not have even been finalized by this point in time. You *will* need to have the software purchased and your environment in place by the end of the module design and configuration session, in order to maintain project momentum.

Typical going-forward tasks are:

- Meet your software account representative
- Finalize your chart of accounts
- Obtain logins for your software system
- Check employees' schedules for the upcoming module setup meetings
- Compile your reporting needs from your existing system and from "side" systems

The outcomes of this kickoff session are:

- A communicated conversion and migration path
- An agreed-upon project scope
- Organizational coordination
- A detailed implementation map
- Reduced anxiety
- Hope

We are now almost done with the planning efforts. In the next chapter, I will discuss the module design and configuration meeting process. After these meetings are finished, the project planning effort will be over. You will be tired at the end of these meetings, and maybe even a little annoyed. I'm telling you right now that I'm going to make you work...

## CHAPTER SUMMARY

1.  *The kickoff meeting is the official start of the software migra-tion and implementation project.*

2.  *The software migration effort will affect everybody in some way and at some point as we move forward together.*

3.  *Some companies elect to have the project sponsor stand up and say a few ceremonial words before we begin presenting the project documents.*

4.  *Your role in the kickoff meeting is to sit, listen, and to* **think**... and to ask questions as they come to you.

5.  *The summary work plan is a one-to-three page, landscape-oriented, summary flow of major tasks and milestones.*

6.  *An issue is a project-related question that needs to either be answered or considered.*

CHAPTER 6

# YOUR SOFTWARE IS A LOADED GUN – PROCEED WITH CAUTION

I'm going to give it to you straight up: The module design and configuration meetings involve long days and they are the most difficult stages of the implementation. Like the discovery meeting and the kickoff meeting, these are on-site meetings between the Lupine consultant and you and your team. Unlike the kickoff meeting, however, not every module design and configuration meeting requires the presence of every team member. The meetings are organized by software module and should include only the people in the company who may be affected by the implementation of that module. For example, the general ledger module design and configuration meetings are almost always attended exclusively by people in the Accounting and Finance departments. Operational personnel are rarely in this meeting category. The meeting schedule is usually presented at the kickoff meeting as a going-forward item. It is NOT unusual for there to be a person or two on the project team who attends every module design and configuration meeting. There is no general rule here–except for this one: If you don't need to be at a meeting, then don't attend. If you do, then do.

Skip holding these meetings at your peril. Yardi Voyager is a loaded gun at this point. You need to be properly trained in its use and capabilities. And you need to have deliberative thought and patience as you go through the process. Your intended organizational use of Yardi Voyager needs to be examined from every possible angle. The module setup meetings should really be called module *decision* meetings. That's the main purpose of the meetings–to make operational decisions about how you want to have the system configured, before you ever even go in and work on the software. It's a purposeful effort to slow you down enough to make decisions, and then, after the meeting, to execute them. We go slower so that we can finish faster.

The most frequent decision points I have witnessed over the years revolve around these topics:

- Chart of accounts
- Departmental accounting
- Electronic funds payment – ACH
- Residential unit type methodology
- Residential amenity based pricing
- Definition of residential traffic
- Commercial recoveries
- Training strategies
- Property roll-out strategies
- Data conversion strategies
- Financial statements
- Custom reports

If you recall, during the discovery session, we talked about the *what* and the *when*. Now we are going to focus on the *how, who,* and *where*. *How* is the setup and migration going to occur, *who* is going to do the work, and under which set of systems (*where*) will the migration and data conversion orchestration take place?

The session will start with our module expert giving you an overview of the module. Typically this overview is projected on to the wall or screen from our laptop. This is not training, and it is not meant to be. It is done to give you some background so we can have meaningful module setup discussions. After the overview, we will begin asking you very specific questions about how you want to set up and configure the system. The good news about Yardi Voyager is that it has terrific flexibility. The bad news is that it has terrific flexibility. *In short, you have to decide how, and if, you want to use this flexibility...and how you want to tie in Yardi's functionality to your unique way of doing business.* Think of it as stretching and configuring the software product to your operational style and methodology. Often, operational processes have been in place for years and sometimes decades. And sometimes these processes are a direct result of the previous software system. The old system may have mandated that a certain process or procedure be handled a certain way. These module design and configuration meetings give you the chance to revisit everything from a brand-new perspective. It is an important, but fatiguing, set of discussions.

The module setup session will end with a discussion of your reporting needs. You will be responsible for compiling all of your reports and bringing them to the session. We will go through each report and put it in one of three groups:

- Match with existing Yardi Report
- Match with existing Yardi Report–but requires modification
- Requires new custom report

We will create a matrix with all of the reports and will deliver back to you an order-of-magnitude estimate of what it will take to create the reports in the 2nd and 3rd "buckets." From there, you can decide whether to proceed with custom report development or to use a combination of existing canned reports to supply you with the information you need to make management decisions.

A module design and configuration meeting typically takes between four and eight hours. The length of the meeting depends on your ability to stay focused on the task at hand, to avoid having side meetings, and to be disciplined. We have had these meetings take as little as two hours and as many as 20. The variables are the three behaviors listed at the beginning of this paragraph. If you are tired and frustrated at the end of these meetings, then you have probably completed a thorough process and properly vetted all relevant issues.

The outcomes of this session are:

- You are now ready to set up and configure their system
- The software will be tailored to your operational methodology
- Operational and software decisions have been documented
- You are more familiar with the system and how it works
- You have completed the implementation planning process

The following pages show typical agendas for the most common module design and configuration meetings. The agenda is just a guideline or map to help us stay on track. The most important part of the sessions is the operational discussions that occur and the system decisions that are made and documented.

## GENERAL LEDGER MODULE DESIGN & CONFIGURATION MEETING

1. Commercial System Review
   a. Dashboard
   b. Property structures
      • Property, building, floor, unit

   c. Lease structures
      • Customer, lease, amendment, unit
      • Recurring charges

2. Global System Configuration
   a. Accounts & Options – Resident Options Tab
   b. Accounts & Options – Receivable Accts Tab

3. Commercial System Configuration
   a. General
   b. Lease types
   c. Recovery setup (related to property/building setup)

      d. Lease clauses

      e. Custom option type

      f. Lease asset type

      g. Custom option types

      h. Lease asset type

4. Deal Configuration
   a. General
   b. Deal probability
   c. Deal stage

5. Roles
   a. Contacts

6. Charge Codes
   a. Required charge codes

7. Property Setup
   a. Properties
      - General
      - Lease types
      - Area label
   b. Units
      - Area measurements
   c. Copy property configuration

8. Leases
   a. Original and amendment lease records
      - Activating
      - Lease terms
      - Options (e.g., renewal, expansions, contractions, terminations)

9. Reporting Requirements

10. Tasks Going Forward

# JOB COST MODULE DESIGN & CONFIGURATION MEETING

1. Overview of the Job Cost Module
   a. Cost Codes
   b. Jobs
   c. Budgets
   d. Payables
   e. Draws

2. System Setup
   a. Cost Codes – Categories
   b. Job IDs
   c. Subcontract IDs
   d. Change order codes
   e. Accounts and Options
      - Environment options
      - Default accounts
      - User defined names
      - Budget options
      - Job 2nd page fields

3. Budgets
   - Workflow decisions
   - Security issues

4. Note about payables and processing checks
   a. Construction payables vs. "normal" payables
   b. Pay retention and required GL accounts

5. Draw Decisions
   a. Types of draws
      - Percent complete
      - Draw from costs
   b. Forms to be used

6. Reports
- c. Yardi vs. Custom
  - What works out of the box
  - What works with modification
  - Custom

## MAINTENANCE AND WORK ORDER DESIGN & CONFIGURATION MEETING

1. Overview of Maintenance System
   - a. Work order dashboard
   - b. Service request vs. work Order
   - c. Work order form
     - Priority
     - Category
     - Sub-category
     - Notes
     - Labor/Equipment
       1. Employees
          - a. Types
          - b. Rates
       2. Start/stop times
     - Material
       1. Tracking stock
     - Custom fields
   - d. Existing reports
   - e. Are work orders be used by all property types (condo, rental and commercial)?

2. Admin Setup Options
   - Accounts
   - Options
   - Employee setup
   - Skills and rates

3. Work Order History

4. Training
   a. How many users to be trained on residential
   b. Full access, read-only, reporting?

5. Reporting Requirements

6. What Else?

An optional implementation activity at this point in time is to engage in "go-live simulation." The go-live simulation phase is the part of the implementation where select end users test and validate the processes and decisions made during the module design & configuration meetings. The purpose of this phase is to provide the end users with the opportunity to discover how the system will handle the newly-defined functionality (as determined by the module design & configuration meetings) and to make any necessary configuration and/or process changes prior to go-live. This phase will take place after the initial setup of the system has been completed. Lupine will give a general overview of the system to the users involved in the simulation testing. These users are typically experienced end users with both policy-based and procedural knowledge of how the current systems work.

All of the testers will need to be in one central location for the go-live simulation process to be effective. The testing/simulation will be performed in a test database environment, which should contain at least one of your active properties, your chart of accounts, and other any other pre-determined data to be tested. This simulation will be the opportunity for the end users to compare the processes in Voyager to the way the existing system handles the same processes, *before* end-user training and go-live. The session attendees should include the Lupine consultant and end users from several different levels of the organization. This will help ensure that the system is being looked at from several different points of view and knowledge levels.

**Note:** There will need to be several different types of individual user account set up in the test environment, so that security settings and permissions can be tested as well.

A go-Live simulation checklist will be provided as a guideline for functionality to be tested. This guideline may need to be adjusted based on your individual company's business practices.

Some key items to test:

- Leasing workflow process (residential or commercial)
- Entering and posting charges, receipts, payables and journal entries
- Summary vs. detail batches
- Prepayments
- Segments (if applicable)
- User security
- Reports

The outcomes of this process are:

- Validation of the decisions made in the module design and configuration meetings
- Accumulation of knowledge of how the system works
- Documentation of the changes that will need to be made to the system
- Assurance that the system is going to work at go-live as intended
- Validation of all necessary end-user training materials

At this point, the planning portion of the implementation is complete. I slowed you down so you could make decisions in a systematic and rational fashion. Going forward, the complete focus of the project will be execution and orchestration. We will now move forward on a week-by-week basis. Each week, you will begin to receive a weekly status report outlining the activities of the project and the responsibilities of each team member. In addition, there will be a status meeting each week with the weekly status report serving as the agenda. Turn the page to learn more in the next chapter.

## *CHAPTER SUMMARY*

1. *Module design and configuration meetings require long days and are the most difficult part of the implementation.*

2. *The meetings are organized by software module and should only include the people in the organization who may be affected by the implementation of that module.*

3. *The main purpose of the meetings is to make operational decisions around how you want to have the system configured prior to your ever going in and working on the software.*

4. *The good news about Yardi Voyager is that it has terrific flexibility. The bad news is that it has terrific flexibility.*

5. *You have to decide how, and if, you want to use this flexibility...and how you want to tie in Yardi's functionality to your unique way of doing business.*

6. *The module design and configuration meetings give you the chance to revisit everything from a brand-new perspective.*

7. *If you are tired and frustrated at the end of these meetings, then you have probably been thorough and properly vetted all relevant issues.*

8. *The go-live simulation phase is the part of the implementation where select end users test and validate the processes and decisions made during the module design & configuration meetings.*

# CHAPTER 7

# COME TOGETHER, RIGHT NOW...

Weekly status meetings play an integral and critical role in the success of your implementation project. The good news is that, if done correctly, they can be completed in approximately 30-60 minutes each week. Limiting these meetings to a maximum of an hour a week allows you the freedom to work on the implementation project as your schedule dictates–understanding that you have a regular full-time job. The status meetings, which usually occur on the same day of the week and at the same time, are held via conference call and led by a Lupine Partners consultant. They begin immediately following the end of the setup meetings, and continue every week until the project is completed.

The agenda for the meeting is the weekly status report that goes out the day before it is held. The report has a summary paragraph, followed by lists of items completed during the previous week, items assigned and not completed, decisions made, and finally, new project tasks that need to be addressed during the upcoming week. The meeting will open with the client talking about any items of concern on the status report. From there, we go through each open task to discuss the status and completion date of the task. It is normal for there to be a fair bit of dialogue surrounding each task. If any one task is getting too much attention, it is an indication that we need to handle that item offline or in a separate meeting. The purpose of the status meetings is what it sounds like: to discuss status.

The weekly meeting attendees should include the client project team members, the Lupine consultants working on the account, and the software account representative assigned to the account. This three-legged stool approach has worked well in having attendees both resolve minor issues and accept responsibility for tasks.

Your job going forward is to complete the tasks assigned to you that week, begin the tasks assigned to you that week, or communicate to us that you will not get the tasks completed. You don't need to think beyond that week's tasks. That's our job. Yours is to:

1. Do the work required by your regular job

2. Fit the implementation tasks (as assigned on the status report) into your normal work day

3. Communicate your status back to us

We will continue in this fashion–week to week to week–until the project is completed. Up to this point the difficulty for you has been in the planning. Now the stress shifts to the "doing." You must manage your time well in order to complete two concurrent activities–your regular job and the Yardi Voyager implementation task assigned to you. The best analogy for this effort is a road construction project. While the new road is being built, the project manager has to keep the traffic flowing even if it means building an ad hoc, temporary driving surface. Same thing here– the business needs to continue operate while your unique Yardi Voyager configuration is being built and tested.

The outcome of the weekly status meetings is that:

- All participants are on the same page
- Problems are communicated and solved
- Everybody continues to work the plan as outlined in the kick-off meeting
- Anxiety is reduced

- Determination is made as to whether the project is on track
- Project camaraderie and professional collegiality increases
- There's a mechanism to stop or delay the project if tasks are not getting completed

The following pages give examples of three real-life status reports: one prepared immediately after the module design and configuration meeting which documents the decisions made during the configuration session, another one written right before the go-dark/go-live dates, and the third report written the week after a successful go-live.

Items to note:

**Status report example 1**. All of the module configurations are noted and the project decisions to date are documented. Now the system build can begin. This is a "speak now or forever hold your peace" moment.

**Status report example 2**. Note the concurrent training and final data conversion efforts. Also note the large number of tasks not completed as time is winding down. This is not unusual. Going live on a new software platform is very analogous to moving into a new home. The house is essentially built but there are 'punch list' items that still need to be completed. Same thing here.

**Status report example 3**. Note that the go-live event occurred and still not everything was completed prior to go-live. Not ideal, but not necessarily a big deal either. Remember that on most implementations, at a minimum on day 1 the following needs to occur:

1. You must be able to bill tenant rent.
2. You must be able to receipt money when it comes in.
3. You must be able to enter invoices.
4. You must be to write checks.

This is the basic blocking-and-tackling, meat-and-potatoes activities that have to, at a minimum, be functional at go-live. Approximately 40 days after go-live, you must be able to publish the first month's financial statements out of the new system. Occasionally, there are some inter-faces that must be in place at go-live. Two that come to mind are tenant payment portals and payroll journal entry interfaces.

At this point you have moved into your new 'home', but there is no grass in the front yard and the custom shelving is not yet up. But – you are able to sleep in your bed at night, not get wet when it rains, flush the toilet, and the Direct TV signal is strong.

Lawson Companies – **Example 1**

Yardi Implementation Project

Status Report for the week ending June 3, 2011

Summary

The contract with Yardi has been signed. Additionally, the decision was made to purchase the Investment Management module and Lawson is currently negotiating the addition of that module with Yardi. Due to the decision to purchase this module we will need to conduct an additional design and configuration meeting on this module. This meeting can be handled via a Go-To-Meeting between Lawson and Lupine.

With the exception of the Investment Management module, the Module design & configuration meetings have all been held. During these meetings decisions were made on how the Yardi modules were to be configured. These decisions are documented below.

Key project dates as decided during the module design & configuration meetings are as follows:

10/25 – Last day to enter data in New Star

10/26-28 – Training on all modules

10/31 – Go Live all modules and interfaces

The accounting team has been diligently working on creating the new chart of accounts. Once that has been completed, we can create a mapping document of the old chart to the new to aid in the import of data from NewStar into Yardi.

Weekly status meeting calls will be held on Tuesday's at 11am EST / 10am CST. The first status meeting is scheduled for Tuesday, June 7th.

Project Tasks Completed This Week

1. Conducted module design & configuration meetings.
2. Signed Yardi contract.
3. Purchased all required servers.
4. Defined property setup.
5. Defined lease options to be setup.
6. Defined tenant variable data.

Tasks Assigned and Not Completed

1. April 21, 2001

    a. Set up Yardi environment (load packages, service packs, and plug-ins). (Lawson)

    b. Send project documents to Yardi account manager. (Lupine)

    c. Hold coordination meeting with Yardi account manager. (Lupine, Yardi, Lawson)

    d. Provide Lupine with appropriate logins. (Lawson)

    e. Complete chart of accounts. (Lawson)

    f. Create property mapping from New Start to Yardi. (Lawson)

    g. Begin vendor 'clean up'. (Lawson)

    h. Create vendor mapping from NewStar to Yardi. (Lawson)

    i. Complete mapping of NewStar cost codes to Yardi chart of categories – for both construction and investments. (Lawson)

2. May 13, 2011

    a. Create GL account mapping between NewStar and Yardi. (Lawson)

b. Review list of jobs to determine which jobs need to be brought over into Yardi. (Lawson)

c. Provide Lupine with approval workflow guidelines. (Lawson)

d. Provide list of users for training. (Lawson)

## Key Decisions Made This Week (Readers: Take Note)

General Ledger:

1. Will be using recurring Journal Entries.

2. Will be using segments and charge codes for the promotion/ retail & mall properties.

3. Properties like Smithville and Lincoln Place will be separate entities.

4. Accounts & Options

   a. Timeout (minutes) - 60

   b. Data Cache – 15

   c. Attachments Extensions: pdf

   d. Attachments Size Limitations: 1000

   e. SQL Reports Prefixes: ler

   f. Default filter rows: 1000

   g. Want the password complexity package loaded: Yes

   h. Shareholders: No check

   i. Manual Property Close: Check

   j. Multiple Properties on JE: Check

   k. Automatic Close Year: Check

   l. Date Warning: Check

   m. Budget Warning: No check

    n.  Display Tran Creator: Check

    o.  Enter Currency Fixed Format: Check

    p.  Show Years with 4 Digits: Check

    q.  Auto Code Vendors: No check

5.  Accounting Books:
    a.  Will set up 13th period accounting books

6.  Will be using user/property security.

7.  Will be changing property codes.

Accounts Payable:

1.  Accounts & Options

    a.  Error for duplicate invoice number = Check

    b.  Invoice number required = Check

    c.  Show detail on the check = yes (want to add property name not property ID)

    d.  Require check approval = No Check

    e.  Prohibit checkbook changes = Check

    f.  Vendor Tax ID required = Check

    g.  Discount property = No check

    h.  PO required for invoice over = Leave blank

2.  Will be changing some of the vendor IDs?

3.  Will you be using blank check stock?

    a.  Stub/check/stub

4.  Will not have signatures printed on the checks.

5.  Will be changing some bank account id's.

6.  Will not be using approvals in Yardi.

7.  Will be using Purchase Orders.

8. Will use intercompany – one central disbursement account.

9. Will use file from bank to reconcile checks.

10. Will not be using positive pay.

11. Will set up a custom table at the property level to list the subscriptions.

12. Will set up Property owners (for the purposes of 1099 owners).

13. Will set up recurring payables.

14. Will not use the invoice approval.

15. Some vendors will be consolidated.

16. Will be using expense types in Purchasing requisitions (testing the PR process).

17. Will set up item types (Lawson part numbers).

18. Will remove the 'approve PO' from the menu – Alan wants the approvers to open the PO and review it to approve it.

19. 1099's will be processed out of Yardi for 2011.

Job Cost:
1. Options:
    a. Show closed jobs = No check
    b. Auto revise budget = Check
    c. Allow contract changes from job = No check
    d. WIP control account = Check
    e. Unit based accounting = TBD
    f. Restrict categories by job type = Check

2. Job Budget:
   a. Budget column = Revised budget
   b. Notification = Warning
   c. Approvals = No Check
3. Format:
   a. Cost category format = ####.##
   b. Max number of characters for the job code= 12
   c. Max number of characters for contract code = 16
   d. Retention distribution = No Check
4. There will be no recurring journal entries in job cost.
5. There will be no purchase orders for job cost.
6. Will change 'Draw Budget' to 'Bank Budget'.
7. Will change 'Bid' to say 'Contract'.
8. Will use cost codes to designate phases.

Investments:
1. Will be using job cost to track investments.
2. Will use job cost code.
3. Category will the name of stocks/bonds (positions)
4. Will set up cost codes (P, I, C)

Investment Management:
1. Will be purchasing the Investment Management module to handle the following:
   a. Consolidating entries
   b. Eliminating entries

    c.  Consolidation Reports

    d.  Investor Rollups

2. All prior decisions made regarding these items are no longer valid due to the decision to purchase the Investment Management module in Yardi.

Commercial:

1. Will be using Buildings and Floors.

2. Will not be using Customers.

3. Will use existing charge codes.

4. Area Labels – Gross Leasable Area

5. What will you use as your "preferred area" column? GLA (Gross Leasable Area)

6. General Tab options:
   a.  Proration type = Monthly
   b.  Default industry class group = NAICS
   c.  One unit per amendment = No
   d.  Require opening charges = Yes
   e.  Manage large unit count = No
   f.  Show lease summary rent = Monthly
   g.  Show general info tab on lease = Yes
   h.  Show other into link on lease = Yes

7. Lease Types will be setup as follows:
   a.  Retail – net
   b.  Retail – fixed

8. The following lease clauses will be setup:
   a.  ADD – Additional Space
   b.  ALL – Construction Allowance

    c. AMD – Amendment Note

    d. Wav – waived rent

9. Customer Types will not be used.

10. Lease Asset Types will not be used.

11. Letter of Credit User Defined Fields:
    a. Bank that LOC is drawn on

    b. # of days' notice required from bank for expiration

    c. original LOC expiration date

    d. final LOC expiration date

    e. Amount

    f. Y/N? – self renewing

    g. Periodic Reductions.

12. Custom Options will be:
    a. Co tenancy

    b. Early termination types

    c. New building relocation

13. Sales Categories will be as follows:

| Catg | Description | Abbrev. |
|---|---|---|
| a. ALT | Alternate Rent | ALT |
| b. ARR | Alternate Rent Relief | ALT RR |

14. Sales Types/Groups will be as follows:

| | Catg | Description | (10 Characters) | Group |
|---|---|---|---|---|
| a. | Catg | Description | (10 Characters) | Group |
| b. | 10-1 | Apparel – Men's | MENSAPPARE | 10 |
| c. | 10-2 | Apparel – Women's | WOMENSAPP | 10 |

15. Commercial Unit Types will be as follows:

| Unit Type | Description | Abbrev. | Unit Category |
|---|---|---|---|
| a. ac | Acres | acres | r |

| b. | Ant | Antenna | Antenna | R |
|---|---|---|---|---|
| c. | Atm | Automatic Teller | ATM | R |
| d. | Café | Cafeteria/Restaurant | Café/Rest | R |

16. Will not be using deals at this time

17. The following Roles will be setup:
    a. Legal
    b. Second Legal Notice
    c. Utility Billing

18. Will be using the same unit numbers.

19. Will be using straight line rents

20. Accounts & Options
    a. Charge notice/eviction = yes
    b. Sales tax on rent = no
    c. Calculate tenant deposits = yes
    d. Update unit rent from resident = n/a
    e. Add'l payor lookup = no
    f. Auto-code Resident and prospects = yes
    g. Default status = future
    h. Prorate rent = yes
    i. Charge move in day = yes
    j. Charge move out day = yes
    k. 30 day month = TBD
    l. Move out transfer = payable
    m. Charge Payment sequence = leave blank

21. Will be setting up the following CPI Indices:
    • CPI-U-67
    • CPI-U-82

22. Historical square footage will not be brought over prior to 2011.

23. Kiosks will be tracked using segments and set up in a separate building so that the square footage doesn't affect cam

24. Notifications will be setup to track the following
    a. Delinquency notices for overdue balances

    b. Delinquency notices for sales report

    c. Upcoming rent increases

    d. Insurance expirations

25. Will setup a custom table at the property level to store audit notes.

26. Will use customers for non-tenant receivables.

27. Will be pulling over the following tenant information:
    a. Pulling in all tenants not expiring prior to 10/31

    b. Kiosk tenants will come over at go live

    c. "Issue" tenants will come over at go live

    d. Square footage history for 2011

    e. All lease iterations for current tenants

    f. Enter in future tenants manually as a training exercise

    g. Past tenants that owe us money

    h. Tenant balances by charge code including prepayments

    i. All recurring charges that have not expired. (no historical recurring charges)

    j. Security deposit balances

    k. All CPI tables with history (see sample)

    l. Three years of sales history (for go-live) then after go-live can bring in any additional history.

    m. Tickler imported into clauses – with dates & type (clause code)

n. Percentage rent data will come over

o. CPI information will come over.

Data Conversion

1. Variable tenant data will be defined as any tenant that has an expiring lease prior to 10/31/2011.

    a. The Following information is being import:

        • 12/30/2010 balance sheet (closed to retained earnings) by property, by chart of account number, by accounting book

        • 12/31/2010 balance sheet (closed to retained earnings) by property, by chart of account number, by accounting book for 13th period entries

        • Monthly G/L detail transactions by property, by chart of account number, by accounting book for January 2011 to October 2011.

        • General Ledger budget totals by property, by chart of account number, by accounting book for 2010.

        • General Ledger budget detail by property, by chart of account number, by accounting book for 2011.

        • Investments (e.g. stocks and bonds) including historical transaction for the codes P & I for all periods.

        • Accounts Payable detail for 2010 and 2011 to the proper G/L accounts.

        • Open payables by property, by vendor as of go-live date.

        • Conversion of Purchase Order History
            1. Length of history TBD after PO test period

        • All active properties/entities and associated units.

        • All current unit square footage information

- All unit square footage history for 2011
- Commercial Tenants
  1. All lease iterations for current tenants
  2. All future tenants
  3. All Past tenants with either a balance owed to Lawson or were billed CAM estimates in 2011
- Tenant balances by charge code (including prepayments for the above tenants.
- All unexpired recurring charges for the above tenants.
- Security deposit balances for the above tenants.
- 3 years sales history for all the above tenants.
- Tickler information will be imported as a lease clause.
- All CPI tables and history.
- All active jobs
- All transaction detail for active jobs
- Job budgets for all active jobs
- Subcontracts for all active jobs
- Interfaces
  1. Commercial Utility Uploads
  2. Electronic banking – cleared checks
  3. Bridge from eSite
  4. Check Scanners (not-Yardi)
  5. BCL
  6. Positive Pay
  7. EDIT Import
  8. Service Orders
9. Payroll (Kronos and current)
10. Daisy

- Lawson personnel will extract the data.
- Lawson personnel will import the data.
- Lawson personnel will validate the data.
- Training
    1. Scheduled for 10/26-28
    2. Lupine personnel will conduct the training.
    3. The training will take place at the Lawson offices
    4. Lawson may use a Train-the-Trainer approach for property, purchasing and payables.

Next Week's Tasks

1. Determine and setup password complexity rules. (Lawson)
2. COA review by Lupine. (Lupine)
3. Format COA for import. (Lupine/Lawson)
4. Confirm that there is a security deposit refund clearing account (required by Yardi). (Lupine/Lawson)
5. Confirm that there is a prepaid rent income account (required by Yardi). (Lupine/Lawson)
6. Add Yardi specific data points to COA (header, total and total-into accounts). (Lawson)
7. Run secondary SQL query to populate additional fields in the ACCT table. (Lawson)
8. Import chart of accounts. (Lawson)
9. Setup accounts & options. (Lawson)
10. Review and setup Commercial System Configuration screen (Setup > Commercial Setup > System Configuration). (Lawson)
11. Review and setup Retail Configuration screen (Setup > Commercial Setup > Retail Configuration). (Lawson)
12. Define and setup Industry Classifications (Setup > Industry Classification Group > Add Group). (Lawson)

13. Create test interface between existing scanners and Yardi.
    (Lawson)

## Lawson Companies – **Example 2**
Yardi Implementation Project
Status Report for the Week Ending October 19, 2011

Summary

Next week we are prepared for a busy schedule. We will begin end-user training on Tuesday, October 25th and it will continue through Friday, October 28th. While training is taking place, Steve and John will be working on exporting the variable data out of the NewStar system and importing the data into Yardi. Once the information is imported, John and Steve will alert the appropriate Lawson personnel that the data is ready to be validated. Once the data is signed off on, the users will be able to begin entering data into the Live Yardi database.

Steve was able to get all kinks worked out of the commercial imports and the commercial data has now been imported into the Live database. The non-issue tenants have been activated.

We have come across some new issues with the Crystal Checks. The main issue lies with the check stub. The stub portion of the check is causing the actual check to not sit correctly on the form depending on how many lines items are pulling into the stub. Both Yardi and Lupine are working to get this issue corrected as quickly as possible. Due to this issue, it was determined that we will use the Yardi Standard check at go-live and up until we can finalize the format of the Crystal Check and get new checks set off to the bank to ensure that they clear.

Particular attention should be paid this week to the Tasks Assigned and Not Completed tasks that are bolded below.

Project Tasks Completed This Week

1. Performed conversion of detailed General Ledger history from 1/1/11 to current for IM entities.
2. Balanced historical data for IM entities.
3. Created eSite interface.
4. Confirmed all Accounts & Options items are setup correctly.

5. Setup commitments.

6. Imported Camrule.

7. Activated 'good' tenants.

8. Created variable data conversion routine.

9. Made first pass at converting variable data from NewStar.

## Tasks Assigned and Not Completed (Readers: Take Note)

1. June 17, 2011
   - Determine how the system will be setup to track the date a check was mailed. (Lupine – 06/17)

3. June 30, 2011
   - Import and test YTD vendor payments into test system. (Lawson – 09/02)

4. July 13, 2011
   - Internal meeting: Agree on and document new AP work-flow. (Lawson/Lupine – 08/26)

   - Setup Requisition>Purchase Order>Accounts Payable Custom Workflow. (Lawson – 09/16)

   - Order and test new check scanners. (Lawson)

5. August 3, 2011
   - Define account trees for transaction rules. (Lawson – 08/12)

   - Test cascade settings. (Lawson – 08/12)

   - Document rollup rules. (Lawson – 08/12)

   - Complete Security Matrix for users. (Lawson – 9/16)

6. August 10, 2011
   - Creation of custom reports. (Lawson/Lupine/Yardi – 10/28)

   - Test positive pay file with bank. (Lawson – 08/26)

   - Create BCL interface. (Lawson – 09/28)

   - Create EDI interface. (Lawson – 09/23)

- Setup users. (Lawson – 09/16)

- Setup property management menus. (Lawson – 09/16)

7. August 17, 2011
   - Run update totals to test that G/L ties to job detail prior to Go-Live. (Lawson – 09/02)

   - Setup check format on all bank accounts. (Lawson – 08/29)

8. August 24, 2011
   - Setup property lists. (Lawson – 09/09)

   - Establish account trees for rollup rules. (Lawson – 09/02)

   - Establish required rollup rules. (Lawson – 09/02)

   - Test settings for rollup rules. (Lawson – 09/02)

   - Document check clearing functionality in Yardi. (Lawson – 09/23)

   - Review lease documents or prior year recoveries detail to define global pools for retail properties. (Lawson – 09/23)

   - Setup system level expense pools for retail properties. (Lawson – 09/23)

   - Review lease documents or prior year recoveries detail to define property level pools for retail properties. (Lawson – 09/23)

   - Assign subset of global expense pools to individual properties for retail properties. (Lawson – 09/23)

9. August 31, 2011
   - Run test check out of system to ensure bank clearing. (Lawson – 10/14)

   - Setup allocation rules. (Lawson – 09/09)

   - Setup equity rollups. (Lawson – 09/09)

Lawson Companies – **Example 3**
Yardi Implementation Project
Status Report for the Week Ending November 2, 2011

Summary

We reached a big milestone this week. We completed the user training and the Lawson team has gone live on the Yardi software.

Training took place the week of October 24th. During that week, Steve and John diligently worked on getting the variable data imported into the system and the rest of the Lawson team validated the data.

While we have gone live on the software, we still have many open items that need to be completed. These items include, but are not limited to: the finalization of the requested custom reports, the completion of the Investment Management setup items, the completion and testing of the remaining interfaces, the setup and training on the Yardi Check Scanners, and the setup of recoveries. John is still working through the import of the paid Accounts Payable items.

Due to the stub/check/stub format, it appears that the Crystal Check functionality is not going to work. The issue being that, in Crystal we have not found a way to tell the report to flow to a second page if the detailed line items on the stub are greater than 4 or 5 lines. The report will simply print the stub continuously and throw the check placement out of line. We have requested that Yardi supply us with a quote for adding the necessary formatting to the Yardi Standard detailed check format.

We still have some hard work ahead of us all, but I would like to thank everyone at the Lawson offices for their hard work and dedication over the last few months. You have all done an incredible job and an incredible amount of work. Your efforts are truly appreciated.

Project Tasks Completed This Week
- Completed the security matrix for users.
- Tested the positive pay file with bank.

- Setup users.

- Setup property management menus.

- Ran update totals to test that G/L ties to job detail prior to Go-Live.

- Setup check format on all bank accounts.

- Ran test check out of system to ensure bank clearing.

- Confirmed all property controls were updated.

- Completed setup of AP system to ensure that it was operational; this included formatting checks for each account - manually or importing into checkfmt and checkfld tables.

- Sent email to users reminding them not use old system at go-dark.

- Setup initial accounting period as the period immediately prior to the go-live month.

- Imported Tenant Balances.

## Tasks Assigned and Not Completed (Readers: Take Note)

1. June 17, 2011
   - Determine how the system will be setup to track the date a check was mailed. (Lupine – 06/17)

2. June 30, 2011
   - Import and test YTD vendor payments into test system. (Lawson – 09/02)

3. July 13, 2011
   - Internal meeting: Agree on and document new AP work-flow. (Lawson/Lupine – 08/26)
   - Setup Requisition>Purchase Order>Accounts Payable Custom Workflow. (Lawson – 09/16)
   - Order and test new check scanners. (Lawson)

4. August 3, 2011
   - Define account trees for transaction rules. (Lawson – 08/12)
   - Test cascade settings. (Lawson – 08/12)
   - Document rollup rules. (Lawson – 08/12)

5. August 10, 2011
   - Creation of custom reports. (Lawson/Lupine/Yardi – 10/28)
   - Create BCL interface. (Lawson – 09/28)
   - Create EDI interface. (Lawson – 09/23)

6. August 24, 2011
   - Setup property lists. (Lawson – 09/09)
   - Establish account trees for rollup rules. (Lawson – 09/02)
   - Establish required rollup rules. (Lawson – 09/02)
   - Test settings for rollup rules. (Lawson – 09/02)
   - Document check clearing functionality in Yardi. (Lawson – 09/23)
   - Review lease documents or prior year recoveries detail to define global pools for retail properties. (Lawson – 09/23)
   - Setup system level expense pools for retail properties. (Lawson – 09/23)
   - Review lease documents or prior year recoveries detail to define property level pools for retail properties. (Lawson – 09/23)
   - Assign subset of global expense pools to individual properties for retail properties. (Lawson – 09/23)

7. August 31, 2011
   - Setup allocation rules. (Lawson – 09/09)
   - Setup equity rollups. (Lawson – 09/09)

8. September 7, 2011
   - Assign allocation rules to entities. (Lawson – 09/16)
   - Assign equity rollup rules to entities. (Lawson – 09/16)
   - Manually setup or import entity lists. (Lawson – 09/16)

9. September 14, 2011
   - Create DAISY interface. (Lawson – 09/30)

10. September 21, 2011
    - Confirm that the 2011 CAM estimate charges and pay-ments have been pulled from NewStar and imported in to Yardi. (Lawson – 09/30)

11. September 28, 2011
    - Setup recovery options at the lease level. (Lawson – 11/11)

12. October 5, 2011
    - Run Entity Commitment Report to display the linear relationships and to audit the setup done to date. (Lawson – 10/21)
    - Run Build Ownership history to show the relationship in a schematic. The stored procedure must be run first. (Lawson – 10/21)

13. October 12, 2011
    - Setup recurring payables. (Lawson – 12/16)

Key Decisions Made This Week

1. Since the Crystal Check format will not work in the stub/check/stub format a request has been forwarded to Yardi to determine if changes can be made to the standard Yardi detail check format and if so, what the time/cost will be.

Next Week's Tasks

1. Import 2012 Budgets. (Lawson – 12/30)

2. Complete documentation for Recoveries Custom Programming Request (CPR) to be sent to Yardi. (Lawson/Lupine – 12/15)

One of the activities we will discuss in almost every status meeting will be the conversion of data from your existing system to Yardi Voyager. And for many, this is the one area that gives them that sick cooties feeling in their gut. Not to fear – we will discuss a number of strategies around the best way to get this data moved over to Yardi Voyager in the next chapter. Stick with me.

## *CHAPTER SUMMARY*

*1. Weekly status meetings play an integral and critical role in the success of your implementation project.*

*2. The meetings are normally held via conference call.*

*3. The agenda for the meeting is the weekly status report that goes out the day before the meeting.*

*4. The meetings attendees should include the client project team, the Lupine consultants working on the account, and the software account representative assigned to your account.*

*5. Your job going forward is to either complete the tasks assigned to you that week, begin the tasks assigned to you that week, or to communicate to us that you will not get the tasks completed.*

*6. You must manage your time to complete two concurrent activities – your regular job and the Yardi Voyager implementation job.*

# CHAPTER 8

# PREPARING FOR A (DATA) MOVE

Now that the module design and configuration meetings are over and the project team is meeting each week, there are going to be three concurrent activities that will have our attention. The three activities are: data conversion, end-user training preparation, and possibly, the creation of any custom reports. In this chapter, we will tackle data conversion.

Data conversion is an area in which many of our clients have had pre-implementation anxiety. To me, it's always been one of the easiest components of the software implementation effort because it's just data. Just a bunch of letters and numbers. Take them from one system and move them into another, like moving into a new home. Now having said that, there is some finesse and planning involved. Here is the mental process (my moving list!) I go through when I start planning the data conversion process for a client:

- Scope
- Cost/benefit
- Methodology
- Source system
- Data cleanliness
- Mapping

- Available tools
- Resources
- Control documents
- Static vs. variable data
- Orchestration
- Validation
- Go-live

**Scope**. Is there organizational agreement as to what should be extracted and converted, and as of what date? What was discussed during the discovery meeting? Is transactional history coming over and, if so, how much? What is the tenant mix–current, future, and past with a balance? Reread and reaffirm the project scope statement.

**Cost/benefit**. This question should be asked: Is there a benefit to pulling certain data? And is it worth the cost? (Cost should be measured in internal employee time and external consulting costs.) For example, if you are in retail and you need gross sales history for ten years, there is a benefit because it affects your rents. You need to think through what it takes to extract data and import it, and more importantly, to validate it. This is the cost that most people don't consider versus the benefit of having the data in their systems. If you need the data, you need it. Period. Discern whether it is a requirement or something that you would just like to have and calculate the cost and benefit.

**Methodology**. Our methodology is simple in its description, but more difficult in its execution. It is this:

1. **Begin the general ledger history extraction, import, and validation immediately**. This strategy assumes that your chart of accounts is ready to go. Converting general ledger history is usually the longest task in the entire project plan. Get started right out of the blocks. How much general ledger history should be converted? There is no right answer here, but it is a big time cost/benefit issue. Is the time it takes to enter and validate all this data worth it? And, more importantly, are

you actually going to use this information? We have had clients who had to move their initial go-live date back because the project moved faster than their ability to convert and validate this data. This can be compounded if you are changing your chart of accounts, because you are validating general ledger account balance information, but the general ledger accounts are no longer the same. What do we see most often? Summarized general ledger history for the current fiscal year and the one prior, so that our client can facilitate current and prior-year comparative reporting.

2. **Convert your static data as soon as possible**. This data is not going to change as a result of the date you choose as your cutover date. Let's get this knocked out so we can check it off our list. More on static data later in the chapter...

3. **Orchestrate the variable data conversion by running go-live simulations prior to the go-live cutover date**. When I was a kid there was a (hippie) saying: "Don't trust anybody over thirty." In the software implementation world, I have my own saying: "Never ever trust that your variable data queries are going to be correct if you wait until the go-dark time period to get started on them." Catchy? No. True? Yes.

Codify the extraction queries and run them against the source system database. Run validation reports out of both systems and check the data. Make corrections in the extraction queries as necessary. Rinse and repeat until you have a match between the two systems. Consider having a "pitching and catching" relationship in which one person is responsible for extracting data from the source system. He or she will "pitch" the data to the person (the "catcher") who will be importing the data into Yardi Voyager.

4. **For go-live, execute the variable data go-live queries**. Besides executing your go-live training program, you should also consider the following:

- Coordinate and use manual systems and tasks during the go-dark process. Transactions still have to be tracked and cash still has to be deposited.

- Compile a list of day one, go-live activities. These activities should be part of the end- user training that occurs during the data conversion/go-dark process.

- Have parties other than the data conversion resources validate the data that has been converted.

- Compile a list of going-forward tasks for the system users after they have gone live on the new system.

**5. Convert non-critical path data items**. Critical path tasks are defined as those that are relevant to making the agreed-upon go live date. If the data is off-critical path, then it means the data elements are not critical to satisfying the go-live requirements. Examples of off-critical path data elements are year-to-date vendor payments that are required for 1099 reporting and filling in data "holes." It's not necessary to have imported year-to-date vendor payments as of the go-live date. The information just needs to be entered or imported in by the January 31$^{st}$ filing date the following year.

**Source system**. What level of access do you have to the source system? For older software systems, it might be none. At a bare minimum, you must be able to print reports to Excel. From there, you can manipulate the data into the import format required by Yardi Voyager. For newer software systems, the chances are good that the back end is a SQL database. If this is so, then you may have the ability to access the tables using query tools. This is the ideal method, because you can create and save queries to use later, when you are going live. And you can write these queries in such a way that the data falls perfectly into the import format required by Yardi.

Yardi has the ability, using their ETL tool, to go natively into several software packages and extract data. Generally speaking, Yardi can electronically convert from any system that has the ability to produce certain source reports directly to Excel. The Excel files are an intermediate step between the initial data extract from the source system and the import of the file into Yardi. Once the data has been saved as an .xls file, then the user can manipulate the extract further. Manipulation examples would include tasks like formatting dates a certain way, adding constants, and deleting NULL values. Once the Excel file is formatted, it will be saved as a .csv file and imported into Yardi.

Should the unconverted data in the old system be archived? You bet– this is particularly true for the data that is not converted to Yardi. Sometimes, you will have access to the legacy system forever. If this is the case, then you can just let the data stay there and access it as the need arises. As time goes on, you will have fewer reasons to go back to the old system. If that is not the case (and especially if you are paying license fees on the old system), then you either need to print hard copies of the reports or save the files in an electronic library folder system so that you can easily go back and find the information.

Does the ability to convert data into any Yardi table exist? Yes—if you can write custom import scripts. To do this, you need knowledge of SQL, Yardi's database schema, and Yardi's scripting methodology. After you have written the custom import scripts once or twice, you will be able to apply that knowledge to importing into any Yardi table.

The last option is to enter all of the information manually. The fact of the matter is that development of the conversion routine takes the same amount of time whether you are converting one property or one hundred properties. Sometimes it is faster and easier to just enter the data by hand. There is no hard-and-fast rule. It depends somewhat on your abilities with data, SQL, and schema. If you are good technically (or you become adept at using the ETL templates), then you will probably be more inclined to go the import route. There is a point, though, at which taking the time to build data conversion templates will save you a lot of time.

**Data cleanliness**. You will have some sort of sense of whether the data in your source system is "good." This is something we always ask and the answer is always "it is very dirty" or "it is very clean." Very rarely do we hear any middle ground in the response. Our clients always know. It seems that the "dirty" data is more common in a residential environment. Perhaps there's been turnover on site and the data just got away from the client. If you have data that is dirty, you must clean it up before you bring it over. You can migrate onto any system in the world, but if your data is dirty then the system is no good. You have gone through this whole process of evaluating software and have chosen Yardi Voyager. And to what end? No software system can help you reach your goals if the data is not pristine.

If the first report that comes out of Yardi Voyager after go-live has un-reliable data in it, then you really haven't moved the ball very far down the field. I have clients say "Well, why don't we just convert and then we can correct the data afterwards?" People in the organization are skeptical and anxious about the whole migration effort from the get-go. You only get about forty-five days after a conversion in which employ-ees will still believe in the new system. Some are even secretly hoping the whole initiative fails so they can go back to the old, "safe" system.

Whereas employees accept your old system as is with bad data, you will not get much of a honeymoon period with your new system. Many people will be quietly skeptical of the organization's ability to actually affect this major change. Here is how the thinking goes: "Bad data, bad decisions, bad software" to "I don't trust the software" to "I will never trust the software." For certain users, all of this can happen in 45 days. Even if you have to push back your go-live date just to clean up the data, do it. This tough decision will pay off for you.

**Mapping**. The next thing to consider is the amount of data matching between the two systems. It's important to realize that your old software system and Yardi Voyager store data differently–they are different sys-tems created by different companies. So, in order to electronically con-vert the data, the fields in the old system must be mapped to the fields in Yardi system. Most users are going to need help doing this, because they are familiar with the schema of the old system, but not necessarily with Yardi's. If you are in charge of mapping the data, it is best to work backwards from Yardi's schema to the source system's schema. This is because Yardi is the "catcher" of the data, while the source system is the "pitcher." Since we are migrating to Yardi, then its schema and data rules are what will dictate the import file structure. Begin with the mandatory columns for a table in Yardi and work backwards to an equivalent column in the source database. If a column is in Yardi, but not in the source database, then this is considered a "gap." Gaps can be ignored, or the relevant data can be compiled outside of the source database and entered/imported into Yardi.

The mapping is normally done in Excel in a From/To or Old/New columnar format. The data elements are represented in the Excel rows. It is highly likely that there will be data "holes" in Yardi Voyager after the data migration. It is also highly likely that Yardi will have more data fields than the source system does. For now, just get the core data in. We'll address the holes later.

**Available tools**. Is this conversion going to be done manually (which means data entry), electronically, or through a method combining the two? There are certain things to consider. The first is the cost of creating an electronic template and methodology to do extracts and imports versus the time cost of just doing it manually. I will give you an extreme example to make this point. Let's assume that the project scope states that you are converting one property, with a balance forward entry for general ledger, with vendors being entered when you enter your first Accounts Payable invoice, and with a commercial property of twenty-five tenants.

This scenario would probably drive you to do a manual data conversion, as you could be finished entering the data by the time you would have finished creating and populating import templates. Another thing to consider, though, is that if you are going to be doing this electronically and if the data is accurate in the source system, it's probably more accurate to import it because all of us make data entry errors. If somebody is sitting there entering in lease information for 8, 16, or 24 hours, he or she is going to make mistakes in the 2%-4% error range. It is going to happen. The data entry person will be interrupted, or he or she will become bored, check email, lose his or her place, and so on. (Or maybe he or she is just a bad typist...)

The three import tools available to you are ETL Templates, Import Scripts, and Import Trans.

Yardi Voyager Extract/Transform/Load (ETL) is a Voyager utility designed to import foreign data from a non-Yardi property management system, or to export Voyager data to a formatted file. The file can then be imported into the non-Yardi property-management system, or into a Voyager database. ETL performs extraction, transformation and load operations to provide seamless data flow between Yardi databases and databases from other property-management systems. It supports data import and export on a daily, monthly, quarterly or yearly basis.

When importing data, you can use ETL templates to associate foreign data-item codes with corresponding Voyager codes (*objects*). You first add a group of ETL objects that specify what to export or import. You then configure a foreign database profile, containing specific options about the data to export or import, as well as other ETL process parameters, including the locations for import and export data files.

*Import mode* consists of a sequential extraction, transformation, and load process:

- **Extraction:** ETL extracts data from a file containing data in either CSV or XML format.

- **Transformation:** ETL transforms the data into logical values that are recognized by the Voyager database-table structure.

- **Load:** ETL loads or imports the transformed data into a Voyager database.

To set up ETL, you must specify where you want the exported data to be saved, where the data to be imported is located, and how you want ETL to transform that data. You specify extraction and load parameters by defining database locations, object lists, and import and export paths. You define the data transformation process by mapping tables and columns between foreign databases and Voyager databases. ETL setup involves three primary procedures:

- Adding groups of objects for ETL to export or import.

- Configuring the ETL foreign-database options.

- Mapping the import data to the corresponding Voyager data codes or values.

The mapping process is completed using one of the following procedures:

Add both foreign-database and Voyager data-item codes and accounts to a CSV mapping file. You then import this information to mapping tables, which automatically display the relationships between the data elements. After mapping, use the mapping-validation procedure to verify that each foreign code or account has a matching Voyager code or account. You can also generate analytical mapping reports to review the details of the operation.

Manually map all your foreign-database codes to Voyager codes. If your import file contains only the non-Voyager data-item codes and accounts (or if the file is not in the specified format), you can add each of the corresponding Voyager codes/accounts to the mapping tables.

*Import Trans* is the function you use to import financial transactions (a journal entry, payable, AP check, tenant charge, or tenant payment). *Import Scripts* is the system function that imports everything *but* financial transactions.

**Resources**. Who is going to be doing the actual work? Will the project depend upon internal or external resources, or a combination of the two? If internal, do the relevant people have the time, and do they have the technical expertise? With the creation of ETL templates, technical capability has become less of an issue. You just need to have decent spreadsheet skills, and basic knowledge of your data.

To create new import scripts, though, you do need some technical abilities. You need to have some familiarity with Transact-SQL, know the Yardi schema, and have some skills and experience in using Yardi scripting. To actually import the files, this isn't necessary. You just need training on how to navigate the import function in Yardi Admin.

**Control documents**. It is very easy to lose your way during a data conversion effort if you do not have a control sheet. There are so many moving parts and you will also probably encounter plenty of interruptions.

I developed this control document methodology and requirement based on hard-earned experience. It's a must. Go to www.lupinepartners.com/ControlDocument.xls for an example. Let's say that you are creating data files, and you have thirty properties. You want to name these files in a way so you can find them easily. If you do run into trouble and you have to re-import files, you want to be able to find them easily. You don't want to get way down the road and ask, "Why didn't I organize this better or differently?"

**Static versus variable data**. Static data is data that, for the most part, doesn't change between now and the go-live date. An example would be units in residential real estate. The tenants change, but generally speaking, the units do not. Static data can be converted at any time during the implementation process—the earlier the better, since time is at a premium as you near the go-dark time frame.

Variable data is data that is unknown until you go dark ("dark" being that brief time period where you don't have a system—for more, see Chapter 11, "Working in the Dark"), and your final data conversion is happening. An example of a piece of variable data would be the name of a residential tenant, and the transactional data associated with that tenant (recurring charges, tenant balance as of the go-dark cutoff, and the tenant's security deposit balance). This data cannot be converted until you close down the old system.

There is one bit of data that is a "tweener," and that data element is commercial tenants. In the residential arena, you cannot determine in advance what the tenant mix is going to be. The move-ins and move-outs are too volatile. But this is not the case with commercial tenants. There is less volatility and you are better able to forecast what that tenant mix is going to be at go-live. Increasingly, we treat commercial tenants as static. We of course keep an eye out for tenants who have a lease end date that is close to the go-live date and note that they may be handled manually at the time of go-live. By and large, treating the commercial tenants as static data has been a good thing for our clients, in that there is less importing to do at go-live when time is tight.

The variable data conversion should be orchestrated and tested for a couple of cutoff periods prior to the go-dark/live process. You want to build a conversion protocol that is executed during the final variable data conversion at go-live. The orchestration process allows you to build a conversion routine that is tested for all conceivable scenarios prior to the go-live data conversion. It is not at all unusual to encounter problems during the orchestration process. In fact, that's why you do it: so that you can find and correct data extraction problems in a safe environment in which you have the time to think through all of the data issues.

**Orchestration**. I have used this term a few times in this chapter already. What does it mean within the context of software implementation and data conversion? It essentially means the creation and testing of extract routines for variable data in advance of the actual go-live date. You want to have performed simulated conversions in advance and have this protocol and methodology ready to go. You have no system, you are down and you want to go grab the variable data and pull it over into Yardi Voyager quickly and accurately. You want to know that it is going to work, and you don't want to have any doubts. Trust me on this; you want it to be 100%. Or, as my friends and I said when we were children, "Shoot low, they're riding Shetlands."

You need to orchestrate the conversion of the data *until it is right*. That could be one time or twenty. In my experience, it usually takes 4-5 times to get the extraction queries correct. Time is of the essence when you are going live, and you have to know that the routine is going to work in the moment. You don't want to be messing around with import scripts when the organization is down for two or three days without a system. Think of the data orchestration process as a dress rehearsal; you just might have several of them before you are ready for a live performance.

Is there any benefit to converting some of the tenant information early in the process, and then bringing in the remainder later on in the process? It's an approach that has some merit. It allows you to enter or import the minimum amount of information required to save a tenant record in Yardi. Once a tenant record is saved in Yardi, the system assigns a tenant ID. When tenant IDs are established, you can go work on other areas of the data conversion that relate to a tenant. It just depends on the implementation plan, and what you are trying to accomplish. For example, if a primary goal of the implementation is the have the work-order history entered early in the process, then you need to have a tenant record established. Later on, you can bring in the less important information. It's a time-saving strategy. Establishing tenant records early (even if you don't have all of the desired tenant information) can free you up to work on other areas of the implementation. The downside of this approach is that you are now doing two imports.

**Validation**. Here's the deal: All of the data that has been converted has to be checked. By people. And it's boring, horrible work. The more people involved in the validation process the better, as different people will find different things. Some people are naturally better at this than others because of their ability to focus and to do detail-oriented work. (Essentially, I'm talking about accountants.) Here are the main reports you should run out of both systems to validate/check the data:

- General ledger trial balance
- Open Accounts Payable
- Rent roll
- Aged A/R delinquency
- Security deposit ledger
- Unit availability

**Go-live**. If the orchestration portion of the data conversion effort has been done well, then the go-live portion can be as simple as executing the queries and then importing the files into Yardi. And that's the point of the simulation efforts. Time is at a premium during the go-live period. There are many things going on and not very much time to get them all done. Plus, we want to keep the dark time down to a minimum. You do NOT want to be figuring out the data mapping and data conversion strategy at this late date. I will further discuss the go-dark and go-live processes in Chapters 10 and 12.

At the start of the chapter, I mentioned the three concurrent activities. Next, I will discuss custom reports, and what they *are* good for…

## *CHAPTER SUMMARY*

1. *Data conversion is an area in which many clients have pre-implementation anxiety.*

2. *Before starting the data conversion, re-confirm the project scope statement with the project team.*

3. *This question should be asked: Is there a benefit to pulling certain data? And is it worth the cost?*

4. *Start the data conversion effort with the general ledger history extraction, import, and validation.*

5. *Convert your static data as soon as possible.*

6. *Orchestrate the variable data conversion by running go-live simulations prior to the go-live cutover date.*

7. *For go-live, execute the variable data go-live queries.*

8. *Convert off-critical path data items after you have gone live on Yardi Voyager.*

9. *To affect an electronic conversion, at the bare minimum you must be able to print reports from the source system to Excel. From there, you can manipulate the data into the import format required by Yardi Voyager.*

10. *Consider archiving un-converted data in the source system–particularly the data that is not converted to Yardi. Sometimes, you will have access to the legacy system forever. If this is the case, then just let the data stay there, and access it as you need it.*

11. *Sometimes, it is faster and easier to enter in the data by hand.*

12. *You can migrate onto any system in the world, but if your data is dirty the new system will be of no use to you.*

13. *It's important to realize your old software system and Yardi Voyager are different systems and thus store data differently.*

14. *It is best to work backwards from Yardi's schema to the source system's schema since the migration is to Yardi. Its schema and data rules are what will dictate the import file structure.*

15. *The three import tools available to you are ETL Templates, Import Scripts, and Import Trans.*

16. *Yardi Voyager Extract/Transform/Load (ETL) is a Voyager utility designed to import foreign data from a non-Yardi property management system, or to export Voyager data to a formatted file.*

17. *Import Trans is what you use to import financial transactions.*

18. *Import Scripts is the system function to import everything except financial transactions.*

19. *It is very easy to lose your way during a data conversion effort if you do not have a control sheet.*

20. *Static data is data that doesn't change between now and the go-live date.*

21. *Variable data is data that is unknown until you go dark and your final data conversion is occurring.*

22. *Data orchestration is the testing of extract routines for variable data in advance of the actual go-live date.*

23. *All of the data that has been converted has to be checked (validated).*

24. *If the orchestration portion of the data conversion effort has been done well, then the go-live portion can be as simple as just executing the queries and then importing the files into Yardi.*

# CHAPTER 9

# CUSTOM REPORTS—WHAT ARE THEY GOOD FOR? (Absolutely nothing...?)

One of the biggest functions of any software system is the ability to deliver informational reports to its users. In general, reports are used to:

- Make decisions

- Present information to customers

- Find errors and omissions in the system

- Balance accounts

- Troubleshoot

- Act as a tickler or reminder system

The reporting capabilities of Yardi Voyager should be important to you. As this concurrent component of the project begins, we need to decide whether the existing reports in Yardi will work for you. My experience has been that some Yardi Voyager reports will satisfy your reporting needs out of the box and some won't. If they don't, then we explore the possibility of creating a unique set of reports just for your organization. These are commonly called *custom reports*.

Here are some of the questions we ask clients when they request that we build a custom report:

- *What is your desired output?*
- What items do you want on the report filter?
- Do you have a mock-up of what you want to see?
- What are the selection criteria?
- What is the report trying to accomplish?
- Do you have a company standard with regard to the heading and other notations?
- What is the desired completion date?

Report specs give the report developer the *Who, What, When, and Where* of the report. The specs basically draw a box around what the user is requesting for a report. Specs are kind of like a scope document for a custom report. The specifications outline the purpose of the report–what you're trying to get done, and any sort of special logic that the report developer needs to take into account. Without the report specifications, the developer is just guessing. Sometimes, even a minor change can require a complete (and expensive!) redo of the report, because of the way it was approached. I've had instances in which I would have developed a report differently had I had all of the desired and required information from the start.

## COST/BENEFIT

Are custom reports really necessary? Do you HAVE to have them? In my opinion, the answer is usually no. Do you need one all-inclusive report, or will a combination of existing (free) reports do the trick? You need to cast an eyeball and do a cost/benefit analysis determining whether the internal and external costs of developing a custom report are worth it. You have a finite amount of time and money. Is the benefit of having a custom report worth the internal and external costs? I have written reports that have cost the client over $5,000, due to their complexity. They could have the same information at no cost by using two canned reports. The goal should be to have all the information required to make decisions—period.

After creating their specifications, I will confirm that I understood the requirements by asking the client if the specs are a good translation of what they are asking for. One of the things we ask clients to do during the module design and configuration meetings is to bring all of the reports they are currently using. These reports could be from their existing system or offline reports, such as side Excel sheets. All these reports are taken and broken into a spreadsheet with three columns:

- There is a match with an existing Yardi canned report (no cost)
- There is a match with an existing Yardi canned report that needs some modifications (cost)
- Or there is no match and there needs to be a custom report (cost)

From there, a time estimate is normally given for each report. The estimate is given to the client, and they decide whether a custom report will be developed. After seeing the amount of effort required to create a particular report, it is not unusual for a client to decide to use one or more canned reports to fulfill the reporting requirements.

## BEING MASTER OF YOUR DOMAIN

If you decide that you do need custom reports, then I would recommend growing or hiring that expertise in-house. You've made a large investment in Yardi, and you need to be able to control your reporting destiny. You must be able to get the data out of the system, in the format and the time frame you want. You don't want to rely forever on outside consultants and the support desk at Yardi. If you do decide to get that expertise in-house, be sure you have two people with this knowledge, not just one. Otherwise, you can get yourself into a "hostage" situation with that particular employee.

Know that in order to write reports in Yardi, you will need to know Transact-SQL at a low-to-mid intermediate level. How hard is it to learn SQL? Answer: Not that hard. The basic elements can be grasped in a day. The most important thing to know is what the Yardi Voyager database schema is. All databases have tables, and within each table there are columns. If you know SQL but not where the data is, then you're not going to be able to develop reports in Yardi. Both the knowledge of SQL and the Yardi schema is attainable in a fairly short period of time. It does help to go to a training class, which will shorten your time frame with both efforts. We have taught many an accountant and/or IT professional to be an expert at this process. Others in your position have done it. You CAN do this.

## GETTING STARTED

When starting a report from scratch, the most important part is making sure the correct data is pulled into the report. I have seen consultants who will actually start at the end and make the report look pretty. That work normally ends up being thrown away, and it is an immature approach to report development. We start in Microsoft Query Analyzer, and create the SQL select statement necessary to pull the data into the report. The data is not going into the Yardi SQL script yet; this is just some meat-and-potatoes hard work to pull the data.

Once that is done, I place the "good" SQL statement within the confines of a Yardi script, conforming to all of the Yardi script rules and requirements. From there, I will integrate a dynamic filter in with the script. (A filter is the mechanism at runtime to decide how you want to limit the report. It's the criteria of the report.) Once I see that the correct data is being pulled and that the report complies with Yardi's requirements for scripting, then, and only then, do I make the report look pretty. Do the cosmetics of the report last. Do the guts of the report first. You'll save a lot of time.

Also realize that sometimes it is easier to have multiple versions of a report with only slight differences among them, rather than having one large report that takes into account all of the functionality of the individual reports. How can this be? Let's say you have a report that is working, we'll call it Report A, and somebody requests that an item be added to the report filter. If this item is to either run the report in edit mode, which only produces a report, or in update mode, which goes into the system and actually updates, for example, the Trans table, then it takes some skill (and time) to actually build this dual functionality. The fastest way would be to keep the original Report A, copy Report A and call it Report B, and make it so the only purpose of Report B is to do the update. Then, there is no dynamic filter item. You run one report to edit, and another one to update.

Creating dynamic filters is doable, but now the report has to be written dynamically to do one of two things at the time you run it, and there can be a fair amount of complexity in doing that. It is actually shorter to create those two reports individually by copying the first report. If time is of the essence, this is how I always handle these kinds of situations. We might go back at a later date to combined the functionalities into one report, but that is up to the client. It may seem counterintuitive, but I have done this many times for clients, because I can give them the functionality while keeping consulting fees down. Using two reports is usually easier (and cheaper) than creating one more intelligent report.

## DIFFERENT STROKES FOR DIFFERENT FOLKS

Is it better to *develop* your reports in Ad Hoc or in Crystal Reports? Actually neither. All Yardi reports start with a SQL script, and these scripts are stored as a .txt file. The script pulls the data from the Yardi tables and sends it to a desired output such as Word, Excel, Crystal Reports or Adobe PDF. You can do further manipulation once you get them into those various products. Even the ad hoc report writer creates a Yardi script. There is a little button you can click which will show you the script. This is how I learned the scripting syntax when I started writing reports in Yardi in 2000–by toggling back and forth between the Yardi scripting user guide and the script that is created in the ad hoc report writer after you click "the little SQL button."

Is it better to *output* reports into Crystal Reports or Excel? There are advantages to both. If you output to Excel, you can further manipulate the data and create macros to automate some tasks. You also get the use of Excel's formulas and formatting, with which most people are familiar. With Crystal, you can't further manipulate the data. This inability to manipulate can be an advantage, since a fair number of our clients don't want the end users to have the ability to change the financial statements.

In Crystal, you also get the use of sub-reports, which gives you the ability to do some pretty interesting programming. And there are formulas conditional of formatting in Crystal just like in Excel. You should know that if you have outputted the report to Crystal, it will not format well to Excel from there. You are better off going directly to Excel if that is your goal or end medium.

If you add a custom table or field to Yardi Voyager, then you can pull that custom data element from the system for inclusion on a report. There just needs to be a correlation between that custom table and the data being pulled. For example: If you created a custom table "below," or attached the unit table to track the size and types of windows in a unit, you would not be able to pull that information into a financial statement, because there is no connection or correlation between the two bits of information–financial data and this custom table created around windows for a unit. But generally speaking, yes—if you have added a new or custom table, then the columns in that table are candidates for use in a custom report. Also, you will need to know the table name and the column name, as you must use that table/column in the actual SQL script.

## DISCRETE DATA

You may have heard about "slicing and dicing" of data. The database needs to be normalized in order to do this. Normalization refers to the concept of breaking data into as many discrete bits of information as possible. For example: there may be five different types of discreet data (attributes) related to the property data you would want to "slice and dice." Typical examples include:

- property type
- region
- owner
- state
- portfolio

If those particular attributes are set up and populated for that property, then you can slice and dice. In other words, it's not so much a reporting mechanism as a database strategy mechanism. Once that is done, you create the report. On the report filter, you would include those five attributes, and then you can begin your slicing and dicing–running reports by property type, region, etc.

Also, it doesn't really matter if you develop your reports on the live database or a test database, because you are just selecting and pulling data. Even if you pull the wrong data on the report, everything is copacetic. It just means the report will have bad data–but the core system will not be harmed. The only time to be concerned about developing a report in the live database would be if you had an Insert or Update statement in the Yardi SQL script. This scenario is possible, depending upon what you are looking to do with the report. It's rare, and it's a more advanced report. In that instance, I would create that report in a test database.

## FIXING YOUR MISTAKES

I am often asked what my process is for debugging errors during the report development process.

There are several different steps you can take after encountering an error message:

- As a first step, I normally take the SQL code into Microsoft Query Analyzer. If you don't get an error, then you know it's not the SQL select statement, and it must be something else.

- Next, you might run the report in debug mode in Yardi to see if anything pops out as being "weird."

- Third, check for any stray commas or apostrophes. Commas and apostrophes (especially the single apostrophe) can cause problems within SQL, because of some special meanings they have within that programming syntax.

- Next, run the report for just one property, and determine if it is a property mix issue.

- If it's a financial statement, I might run the report for just one GL account. I would set up an account tree with only one account. Error? Yes or no.

- After that, begin to take out portions of the select statement and put them back in one by one until you can identify the break.

The bottom line with debugging reports is that you should always, always change just one thing at a time so you can identify the problem when it occurs. If you are changing two or three variables at a time and the report begins to run properly, you won't know which element was causing the problem. It's a very methodical process. I normally turn off the phone and email because I don't want that mental string broken when I'm in debug mode. The last line of defense is to ask somebody else to look at the code. You've been too close to it, and the answer is probably staring you right in the face. More often than not, they will see the error immediately. I have been on both sides of this–I've found someone else's error, and someone has also found the error for me.

Realize that a custom report will almost always work just fine after an upgrade. There are a few instances in which it might not. One instance is if there is a total rewrite of a module, such as Yardi's upgrade from 5.0 to 6.0 for commercial. That upgrade was a complete schema change. Custom reports that worked for 5.0 did not work for 6.0. They wouldn't necessarily "error out," but they certainly did not return the correct information. Another instance to consider is when there are some minor functionality changes—like how traffic used to be handled in residential for previous versions of Voyager. There were some new tables set up around prospects, so Box Score reports were no longer valid. In general though, you will be fine. Ninety-nine percent of the time, a custom report will continue to work as Yardi adds service packs and plug-ins, and you will know in advance when major upgrades are coming along that will change the database schema.

I am sometimes asked if there is one large library somewhere of all the reports that have been developed in Yardi. The answer is no—not even within the Yardi organization. You will have reports loaded onto your default path, but these do not represent all of the reports that exist in the Yardi universe. There are tens of thousands of reports that have been created–by Yardi personnel, by independent consultants, and by Yardi users. Joining a regional user group is a good way to get your hands on some other reports. User group members have been known to share reports with one another.

Meanwhile back at the ranch, we are ready to move on to the third of the three post-planning activities: end-user training. Stay with me, and I'll see you at the next chapter.

## *CHAPTER SUMMARY*

1. *One of the biggest uses of any software system is the ability to deliver informational reports to its users.*

2. *You need to decide whether the existing reports in Yardi Voyager will work for you.*

3. *Report specs give the report developer the* who, what, when, and where *of the report.*

4. *The goal of reporting should be to have the information necessary to make decisions.*

5. *After seeing the amount of effort required to create a particular report, a client may decide to use one or more canned reports to fulfill the reporting requirements.*

6. *If you decide that you do need custom reports, then it is recommended that you grow or hire that expertise in-house.*

7. *In order to write reports in Yardi, you will need to know Transact-SQL at a low-to-mid intermediate level.*

8. *Don't work on the cosmetics of the report until you have gotten the report to function.*

9. *All Yardi reports start with a SQL script, and these scripts are stored as a .txt file. The script pulls the data from the Yardi tables and sends it to a desired output such as Word, Excel, Crystal Reports, or Adobe PDF.*

10. *If you add a custom table or field to Yardi Voyager, you can report off of that custom table.*

11. *Data normalization refers to the concept of breaking data into as many discrete bits of information as possible.*

12. *When debugging reports, always, always change just one thing at a time so you can identify the problem where it occurs.*

# CHAPTER 10

# IF A TREE FALLS IN THE FOREST...

If a tree falls in the forest and nobody hears it, did it make a sound? Or in our case, if a software implementation occurs and people do not how to use the software to do their jobs, did anything really happen–except for an extraordinary waste of time?

The last component of the post-planning activity trifecta is end-user training planning. Question: What is the cost of NOT being trained properly to use Yardi Voyager? Answer: You run the risk of the entire implementation effort being all for naught if the users cannot use the system in the manner in which it was intended. Any custom reports that have been created, the entire data conversion effort, and the setup of the individual modules will be a throwaway if the users cannot use the system. You run the risk of the entire implementation being deemed a failure. I've seen this happen. Generally you have an approximately 45-day window for the user base to say either: 1) they love the software or 2) they hate it. Once they've made that emotional decision, it is tough to change their minds. The bottom line here is: don't go cheap on the training. It is the last piece of the entire implementation effort, but if it is done poorly, then everything that preceded it will probably have been a waste. I'm talking about end-user, go-live training here, not module setup training. The purpose of the module setup training, which happens at the beginning of the implementation, is to demonstrate the software to the extent necessary to make decisions around the intended use of the particular software module. (See Chapter 6 for more information on the module design & configuration meetings.)

## DOING TWO THINGS AT ONCE

The go-live training should ideally occur at the same time the final data is being converted, aka the 'go-dark' period. Go-dark, which will be discussed in detail in the next chapter, is the period after you have taken down the old system and right before you go live with the new one. It is typically a two-to-three day period, in which you are system-less while your data is being converted from the old system to the new. This "dark" time period that is the ideal time to train your users. When the training is over, the data from the old system has been converted and everything is ready to go. The users can then begin going through the list of items and tasks that need to be covered while they were off at training. (For those two to three days without a system, any transaction that occurred will need to go into Yardi Voyager.) User training should be completed as close as possible to the go-live timeframe. If it is more than two weeks before go-live, then they're probably going to forget what they learned in training and you run the risk of having to retrain them.

This concurrent effort of training and converting the data accomplishes two things. One, it maximizes resource usage: the people doing the data conversion are not the same as the ones who are leading the training effort, so you can have simultaneous efforts. Two, it reduces the amount of time for end users to forget what they learned. Ideally the system users go to training, get out of training, and then go live with Yardi Voyager. If they go right into using the system (and not go back to the old system), then everything they learned is fresh in their minds.

## HANDS-ON

The class sizes should be anywhere from one to ten people. Once the number goes above ten, even the most disciplined groups will break down and begin having side conversations—at which point they are obviously not listening to the trainer. The trainer is now training two to three different groups, all of whom are engaged various conversations.

The smaller the class, the higher the focus of the attendees. If you want to train more than ten people, then I recommend that you break it into two or more sessions. Even though it is more expensive, I would always lean towards having the smaller group—more learning will occur. If you go through the process and the system users do not learn the software, it increases the possibility of a failed implementation. Also, each trainee should have his or her own computer for the training session. With software training, it's impossible for users to just watch what is being done. And sharing doesn't really work either. One person in the pair will become the teacher and the other the student, while they both need to be students. It is just not a good dynamic to pair up and "watch" a software training session. You must have trainees' hands on keyboards...

You might want to contemplate using Quick Guides or Student Aids. Even though Yardi provides user guides, the fact of the matter is that most people won't use them. It's not because they're not well-written, but rather that they're thick and daunting and most people want a quick guide to show them how to do what it is they are going to be doing day after day after day. People will use index cards or one-page documents that show them how to do repetitive tasks. Part of your training implementation program should include the creation of these guides and/or aids. They can be passed out during the training session, after each training module, or at the end of the entire training process. When they go back to their desks after going through the Voyager training, they will have these guides available to use during the first few weeks. After that, most of the repetitive tasks have been memorized and the materials will no longer be needed. (But they can be used by new employees–*don't throw them away!*)

## HIS DATA, YOUR DATA, OR MY DATA?

A common training question revolves around what data should be used for the training. Ideally, you should use your own, but practically speaking this is difficult to do and it's usually not cost-effective. This is because *your* database has to be set up to be meaningful in a training environment. Sample or training data has usually already been set up to demonstrate the software functionality, whereas your data will not typically have been set up this way. "Your" data is usually the remnants of the data conversion orchestration effort. *What is the cost or benefit of setting up your database for training just so the users will recognize the properties, units, and tenants?* Yes, they are more familiar with the data, but does it actually help them learn the software faster or better? My thought is no....you can spend your finite time and resources in other areas on the implementation. Just use a training or sample database.

Should you put beginner and advanced users together in a training session? All of us have had experience with this over the years, either being the smart, experienced person who is bored because there are newbies in the class, or the inexperienced one, with everything going over your head. And it's the same with Yardi; if you make the mistake of combining groups with varying levels of experience or interest, then you run the risk that one side will be bored and the other overwhelmed. This is a question to ask and analysis to do in advance, because your goal is to get everybody trained. You may think you are saving money by putting everyone in the same class, but if only half the class is actually being trained, you have not furthered your goals of getting everyone trained on Yardi Voyager. You didn't really save any money if one group did not get what they need.

## FACE TO FACE

The final end-user training must be done in person. You have to be able see students' reactions and you must be able to monitor the room to see what's going on. They also must be able to see your face as you're instructing them in the use of the product. Even with new technology, there are some things that still must be done face-to-face to communicate the material. Remote training can be used to work through particular software or user issues, but it's not a full end-user training. Don't try to save money by having the instructor teach remotely.

What about listening in by phone? This does not work. Period. The people calling in will be checking email and doing work for their regular jobs. No training will have occurred. You are better off having them not attend at all. If they attend the training by phone, then in their minds they will have been trained. There may be a check box marked off by their names, but they will not actually have the required software knowledge. Resist the temptation to save money on training. If you want to cut corners, do it elsewhere (like custom reports). Invest as much money as you can here–have more sessions, not fewer. Have smaller class sizes, not bigger. Be smart about this.

Here are some training rules we regularly use at Lupine while conducting training:

- Write down your questions. Do not interrupt the instructor as he or she is going through the training. There are other people in the room besides you. We will get to your questions at the appropriate time. (Discussions on the use of a "parking lot" are forthcoming...)

- No side conversations (meaning, with your neighbor). Besides being rude, it prevents the person you are talking to from being able to hear the training. Side conversations are extremely disruptive.

- No cell phones, for what I hope are obvious reasons: They are disruptive to you and the entire class when they ring. And there is no better way to ensure your own failure than to spend your training time texting. You can return calls during break.

- No email. You need to focus on what the instructor is saying, not on managing your inbox.

- Multiple breaks. This is an intense process. I usually run the class for 50 minutes at a time with a ten-minute break and then go on. This leads right me into the next point...

- Start and stop on time. If you're going to start at ten, then start at ten. When I run a session that starts at ten, I start talking at ten o'clock. You have to keep these things moving at a pretty brisk pace.

I see most training sessions break down due to a lack of discipline. These sessions are short bursts of concentrated activity. One way to ensure that everybody maintains focus is to tell the users that they will be tested at the end of the training. Then test them. Having a quizzing policy is important. Think back to when you were in school. If you knew for a fact that a quiz was coming, wouldn't you focus more during class? Wouldn't you therefore retain more information? Maybe learn more?

You are investing a lot of money in this implementation process. I submit to you that investing money in an end-of-training quiz is worth it. By doing so, you may discover that you have a less-than-stellar trainer or you may learn that certain people from your side did not apply themselves. You could discover that some individuals may not be competent enough to use the software; you may want to "hold back" these individuals and make them take the class again, or even show them the door. It is better to learn this news sooner rather than later. Yes?

## PAVED PARADISE... PUT UP A PARKING LOT

We use a technique called a "parking lot" to help maintain order during a training session. A parking lot is a training tool to temporarily park a question that, while important or relevant, is not so at that particular moment. With the proper use of this tool, the training can continue on pace and not get sidetracked by a well-meaning question. The valid comment is saved for later and addressed down the line. Typically, the "parking lot" is a white board and I assign somebody (other than myself) the job of actually standing up and writing each item down so that everybody sees that his or her valid comment will be captured and valued. The dynamic that begins to develop is that training attendees don't mind that their question is not being answered immediately because they know it's in the parking lot. At the appropriate time, all issues/questions that have been "parked" will be addressed and discussed by the entire group. If you use a parking lot for your training sessions, then you will have much more disciplined classes. You now have a respectful methodology for dealing with people who ask questions that are not on point or premature in the training process. I usually ask permission at the beginning of each session to use a parking lot.

Also, there is no upside to having Internet connectivity during the training session. There are no pros, only cons to making it available. If you give people access to the Internet, they will surf the web, check email, instant-message with their friends, read the news, and look at their stock quotes. You get the idea. Don't turn on the Internet unless you need it for the training session itself. There is no other reason to do so.

## WHO CONDUCTS THE TRAINING?

One of the early tactical decisions to make is whether you want an outside subject matter expert to train your end users, or whether you want to "train the trainer." This process happens when the subject matter expert comes in and trains people within your organization to give them the software product knowledge. Then the trained people from within the organization will go on to train the remaining members of your organization. This is an alternative to trainers coming in and directly teaching people in your organization. In the "train the trainer" model, the internal trainers serve as kind of a middleman in this process. They obtain the software knowledge and then go from there. The "train the trainer" approach can work very well if you have:

- A training department
- Many, many users to train or properties to roll out
- An employee who is naturally adept at training or software, and has a strong urge to lead this training effort, and has the time to do it
- An employee who has past experience training others on or using Yardi Voyager

Otherwise, you are probably better off hiring out the end-user training process. If you do go the "train the trainer" direction, it helps to have the internal trainer on the project team from the start. If not, he or she will have to be brought up to speed as to how the Voyager module was configured and other relevant items that only the project team would know. The training could have bad outcomes if the trainer and the project team are not in sync. This doesn't have to be a big deal, but a coordination meeting should be held to discuss the status of the implementation, the training dates, the training agenda, the special needs and nuances of the people being trained, and things you DON'T want to be trained on.

For all the trainings, whether they are end-user or train the trainer, it is ideal to procure a training center for the session. You're going to be able to focus more: no phone calls or emails, and no boss coming along to disrupt the flow of the session. This will most likely cost more because you have to rent the room, but it's quite possibly worth it considering the cost of software failure. Also, as soon as the training dates are assigned (which is usually done during the module design and configuration meetings), you can go ahead and reserve the training facilities (even if it is your own conference room) and get the trainers scheduled. Get a slot on the trainers' calendar so another company doesn't grab those dates. If you do it early, then you can check off that box and know that you will have the professional resources you need.

## IF IT'S CANNED – CAN IT!

I should have mentioned this at the beginning: Do not accept a canned training. This means you are going to have to be proactive with your trainer. I recommend that you to go ahead and create a first draft agenda and send it to your assigned trainer. There will be some back and forth between you and the trainer as you strive for clarity. If you do not do this, then you are going to get a canned agenda that will almost assuredly not suit your needs. The trainer is not going to know what your hot spots or pressure points are, or what your weaknesses are. There is no way for him or her to know. Do not abdicate this responsibility. You want an agenda that has been customized to your needs and desires.

There is a benefit to the trainer having a copy of the database prior to the training: he or she can see how your system has been configured and can have his or her training database (if he or she isn't using yours) configured the same way. Having the trainers look at the database in advance ensures that the functionality the users are trained on is the same one they will see when they get back to their desks and start using the live new system. Don't skip this step. It is very important for the trainer to know your configuration. If he or she doesn't care to see it, it means you will probably get a canned training that won't suit your needs. Get yourself another trainer.

## LET'S GO TO THE VIDEO

Another training tactic to consider is to record the entire training program to video in advance of the first user training. In recent years, we have seen more and more of this. The benefits to this video approach are:

- It costs a lot less than a traditional training
- You have training materials that can be put into a library for future use
- The Yardi Voyager knowledge base will not reside with a particular person but rather within electronic files
- A quizzing component can be added to your training strategy
- Your new hires can be trained quickly without a large trainer mobilization effort
- The series could be used as a hiring tool before an offer is extended
- You don't need to be an expert in Voyager to facilitate the sessions
- Multiple people can serve as client facilitators
- Travel costs can be reduced or eliminated.
- Training is uniform—everybody receives the same level of training. Nothing is left out and nothing is added.
- The trainings can be repeated. Clients can have them day after day after day. Unlike a human trainer, the video never gets tired.
- You have flexibility with regard to the location and timing of the training.
- Copies of the video can be made and used at various sites.

The videos should be a combination of the relevant Voyager screens, accompanied by the voice of a narrator. The narrator will conduct the training exactly the same way he or she would if he or she was in the room with the trainees. Just as for an on-site training, a screen projector will be necessary for the video training. Similarly, each trainee should have his or her own computer.

The key to the video route is that tools are created in advance for the client facilitator to use. The facilitator will receive detailed instructions in a facilitator guide on how to conduct every single second of the session. The scripted guide works in concert with the video, telling the facilitator when to pause the recording. Once the video is paused, the facilitator will guide the trainees through an exercise that pertains to the topic they've just viewed. An exercise will be passed out to the attendees and the facilitator will "walk the room" and help as needed. Once that topic has been covered, the facilitator goes back to the guide, pushes "Play," and waits for the next "Pause" signal. Then the next exercise is covered. This start/pause routine will continue until the Yardi module has been fully covered. At the end of the training, each user should be given a final quiz. You may choose to have a policy that an attendee will not receive his or her full login information until he or she passes the final exam.

This is a worthwhile investment of time. But you should know that this is more difficult than just doing a "normal" training. The payoff comes with the trainings following the first one, and when you have new employees in your organization who will be using Yardi Voyager.

## MAKING A LIST AND CHECKING IT TWICE

Here is a checklist that all of us at Lupine Partners use when we have to conduct a training. It is the byproduct of the hundreds of end-user training sessions we've done…in which we've had many things go wrong. It is updated constantly.

**Class Title:**
- Trainer
- Date of Class
- Time of Class (Including Time Zone)
- Name/Contact Info of Company Contact
- Number of Class Participants
- Location of Training: On-site, Off-site or Remote

- If the location is off site, has a project team member visited and assessed the site?

**Technical:**

- What is the Yardi version number and plug-in version numbers? Are there any known issues?
- Is there someone who will be available on the site if there are any technical problems?
- Will a projector be available?
- Number of computers
- Is there a printer available?
- Has Crystal Reports Viewer been installed?
- Keyboards, mice/mouse pads: available and working?
- Is there Internet access for all?
- Wired access or wireless?
- If wireless, is there a password required?
- If wired, are there enough cables?
- If using laptops, are there electrical outlets and power cords available?
- Which Yardi database will be used?
- What is the URL for logging in?
- Will everyone have access to this database?
- Will everyone's user access be the same?

**Agenda:**

- Date original agenda sent to client
  (attach email confirmation)
- Date final agenda approved by client
  (attach email confirmation)

## Housekeeping:

- Who is handling lunch/refreshments for the group?
- If the above person is not the contact person, has he/she been provided with list of attendees?
- How many breaks should be worked into the schedule? How often should they be?
- List of the attendees for the class, along with their titles
- Who is responsible for the attendee reminder about the class?
- Are there any special instructions before class starts? (Cell phones, restrooms, etc.)

## Materials:

- Are the agendas printed for the class?
- Are there any additional materials needed: flip chart, hand-outs, etc.?

Time to keep on keepin' on. The next chapter is going to discuss the strategy, choreography, and orchestration of the go-dark/go-live process. When properly planned, this can be a beautiful execution of what most think is a horrible process fraught with fear. Or it can be a Chinese Fire Drill—your choice. Read on.

## CHAPTER SUMMARY

1. The last component of the post-planning activity is end-user training planning.

2. You run the risk of the entire implementation effort being all for naught if the users cannot use the system in the manner it was intended.

3. Don't go cheap on the training.

4. The go-live training, ideally, should occur at the same time the final data is being converted.

5. User training should be as close as possible to the go-live timeframe. If it is more than two weeks before go-live, then they're probably going to forget what they learned in training.

6. This concurrent effort of training at the same time of converting the data accomplishes two things. One, it maximizes resource usage—the people doing the data conversion are not the same ones leading the training effort, so you can have simultaneous efforts. Two, it reduces the amount of time in which end users can forget what they've learned.

7. The class size should be between one and ten people.

8. Each trainee should have his or her own computer for the training session.

9. You might want to contemplate using Quick Guides or Student Aids.

10. Think through whether you should you put beginner and advanced users together in a training session.

11. The final end-user training must be done in person.

12. One way to ensure that everybody maintains focus is to tell the users that they will be tested at the end of the training.

13. *A "parking lot" is a tool that allows you to "park" a question temporarily if it's not relevant at that particular moment.*

14. *There is no upside to having Internet connectivity during the training session.*

15. *One of the early tactical decisions to make is whether you are going to want an outside subject matter expert to train your end users, or whether you want to "train the trainer."*

16. *If you do go the "train the trainer" route, it is beneficial to have this person on the project team from the start.*

17. *It is ideal to procure a training center for the training.*

18. *Do not accept a canned training agenda from your trainer.*

19. *By having the database in advance, the trainer can insure that the functionality the users are trained on is the same one they will see when they get back to their desks and start using the live system.*

20. *Consider using a checklist like the one provided at the end of this chapter.*

# CHAPTER 11

# WORKING IN THE DARK

All of your work efforts—the discovery meeting, kickoff and module design meetings, weeks of status meetings, data conversion orchestration efforts and the creation of a terrific training program–have been choreographed into the execution of the final implementation effort, which we call "go dark." This is the point at which you have turned off your existing source system and will temporarily not have a system while your data is being converted from your source system to Yardi. Ideally, the go-dark time period should be kept as brief as possible. With Lupine Partners as your implementation partner, the go-dark period is typically 2-3 days, depending on the number of properties being converted.

As discussed in the previous chapter, the go-dark period is the ideal time to train end users on the modules being converted. When the end users get out of training, their data should be converted and ready to be validated. They can then begin using the training they received to process the transactions that occurred while they were in training on the new system. This post-training, post-conversion period is called "go live." You are now on your new system and ready to begin processing.

The goal of the go-dark/go-live time period is to decrease the downtime in which you have no system, and also to decrease the amount of time between when the end users are trained and when they begin using the new system. All of these critical processes can, and should, be staged in advance. It is a carefully choreographed dance that should not just happen when the go-dark/go-live occurs. Dress rehearsals should be held in order to minimize the go-dark downtime and maximize the success of the data conversion effort.

Our suggested go-dark/go-live approach is to:

- Have a concurrent approach of data conversion/training to use the downtime efficiently

- Coordinate and use manual systems and tasks during the go-dark process. Transactions still have to be tracked and cash still has to be deposited.

- Compile a list of Day One, go-live activities. These activities should be part of the end-user training that occurs during the data conversion/go-dark process.

- Have people, other than those working on the actual data conversion, validate the data that has been converted

- Compile a list of tasks going forward for the system users after they have gone live on the new system

I will admit at this point that I told you a little white lie at the end of Chapter 6. If you recall, I told you that we were through with planning. Oops! There is one more planning meeting and it's a biggie. Approximately 45 days before your scheduled go-live date, we will have a key strategy meeting with you where we will choreograph what the last 5-10 days before you go live will look like. Tactics will be discussed and responsibilities assigned. All of these tasks will be incorporated into the weekly status report for constant reminders during the 4-5 weeks leading up to the go-dark/go-live period.

# The working document for this strategy meeting will look something like this:

## SAMPLE GO-LIVE RESPONSIBILITY PLANNING MATRIX

| | Prior to Week 1 (9/6 - 9/23) | Operations | IT | Accounting | Sites |
|---|---|---|---|---|---|
| 1 | Group planning meeting with regional managers and accountants | O | O | O | O |
| 2 | Check delinquency and prepaid balances. Also check security deposit balances | | | | O |
| 3 | Define source balancing reports | | | O | O |
| 4 | Create envelopes (or other documents) with the go-live checklist on front. These will be used for cash receipts, security deposit refunds, and work orders to be processed in Yardi after the "dark" period. See Excel tab for examples | O | | | |
| 5 | Create procedures to be implemented by onsite personnel during go-dark period. See Excel tab for examples. | O | | | |
| 6 | Email from senior personnel to all sites; roll out schedule and expectations | O | | | |
| 7 | Print any historical reports that need to be archived | | | | O |
| 8 | Process security deposit refunds before go-dark period | O | | | O |
| | **Week 1 (9/24 to 9/28) Go-Dark Period** | | | | |
| 9 | AMSI PM module goes dark at 3:00pm on 9/25/2007 (at the latest) | | O | | O |
| 10 | Run balancing source reports from AMSI (deposits and balance forwards, revenue) - Detailed Rent Roll | | O | | |
| 11 | At sites: make copies of cash receipts | | | | O |
| 12 | At sites: make copies of traffic/application, etc. | | | | O |
| 13 | Monday to Wednesday: HQ training | | | O | |
| 14 | Wednesday to Friday: Site training | | | | O |
| 15 | Wednesday: Create import files | | O | | |
| 16 | Wednesday: Print out relevant balancing reports from AMSI | O | | | |
| 17 | Wednesday: Send read only access of AMSI back to sites (if applicable) | | O | | |
| 18 | Wednesday: Import data into Yardi | | O | | |
| 19 | Wednesday: Print all resident ledgers (this can be done anytime during the week) | | O | | |
| 20 | Thursday/Friday: Balancing | O | ? | ? | ? |
| 21 | Friday: Begin entering demographic information as you have it | | | | O |
| 22 | Make journal entry to back out effects of entering beginning tenant balances, delinquencies and prepayments. Make sure that AR detail ties to GL | | | O | |
| 23 | Close September 2007; turn calendar to October 2007 | | | O | |
| 24 | Send out email to Regionals/Accounting re status of property | | O | O | |
| 25 | Put relevant residents back on notice. Verify availability report | O | O | | |
| | **Week 2 (10/1 to 10/5) Go Live** | | | | |
| 26 | Continue balancing, if necessary | | O | | |
| 27 | The accounting entry should go to fiscal 10/07 | | | | O |
| 28 | Enter roommates (manually) | | | | O |
| 29 | Enter cash receipts from previous week | | | | O |
| 30 | Enter traffic from previous week | | | | O |
| 31 | Enter move-ins and move-outs | | | | O |
| 32 | Wednesday: Verify leasing agents on the property | | | | O |
| 33 | Charge rents, if applicable | | | | O |
| 34 | Manually enter "Additional Occupants" captured during data cleanup. | | | | O |
| 35 | In general, implement Day One go-live plan | O | | | O |
| 36 | Implement post go-live data checklist | | O | | |

## SAMPLE GO-DARK/GO-LIVE PLANNING CALENDAR

| | Sun. | Mon. | Tues. | Wed. | Thurs. | Fri. | Sat. |
|---|---|---|---|---|---|---|---|
| Month 1<br>Week 1 | | | Execute Go Live Minus 3 Tasks | | | | |
| Month 1<br>Week 2 | | | Execute Go Live Minus 2 Tasks | | | | |
| Month 1<br>Week 3 | | | Execute Go Live Minus 1 Task | | | | |
| Month 1<br>Week 4 | | | Shut Down End of Day | Convert Data / User Training | | | |
| Month 2<br>Week 1 | | Implement Go-Live Procedures | | | | | |

Items we typically cover in this important planning session include:

**Source system data cutoff**. What exact date and time will the source system be frozen, locked, and turned off so that users cannot access the data? Who will be responsible for communicating this to all relevant parties?

**Re-confirmation of the data conversion approach**. We usually go through the entire fixed and variable data conversion discussion again. Many times we will email the data conversion control schedule to the project team as part of the meeting. We don't move forward until everybody is nodding their heads up and down to say, "Yes, David, we agree with you and we understand." I will even say, "Is everybody crystal-clear on this? Any questions at all? Speak now or forever hold your peace."

**Go-live data validation approach**. Who will be the validators? What will their approach be to checking the data? Will they be comparing reports one by one? Or will they use a "pitcher/catcher" methodology in which one person reads from the source system and the other checks off the data on the new system reports? There must be clarity on the approach and which people will be assigned to which properties, and which specific data are they validating.

**Status of data scrub (if relevant)**. Are we done, on track, or behind? In my experience the answer to this question is either "done" or "behind." If the answer is "behind," then there needs to be a herculean effort to finish before the go-live date. Otherwise, we are going to have to push back the go-live date. As previously discussed, there is no upside to bringing over tainted data.

**Confirmation of the training dates for each module**. Even though this was set early in the process, we make sure to ask: Are these the correct training days for each module?

**Confirmation of training locations**. Do we know where the training sessions are going to be held? Is the training facility or conference room booked? Even if so, confirm it again.

**Confirmation of the training attendees**. Who is being trained when? This is a good time to have the discussion (again) about the number of people to be trained. Think about capping the sessions at 10 attendees. Have the attendees been notified? Do they have backup for the days they are going to be trained?

**Confirmation of training and resources, both human and technical**. Which trainers will be conducting the training? Confirm that they your training is still on their schedule. Technical items (from Chapter 10) to consider:

- What is the Yardi version number and plug-in version? Are there any known issues?
- Is there someone who will be available on-site if there are any technical problems?
- Is a projector available?
- What are the number of computers that will be required?
- Is there a printer available?
- Has the Crystal Reports viewer been installed?
- Are the keyboards, mice, and mouse pads available and working?
- If the Internet is going to be used, is the connectivity working?
- Will the Internet access be wired or wireless? If wireless, is the password available? If wired, are there enough cables?
- Which Yardi database will be used?
- What is the URL for logging in to Yardi?
- Will everyone have access to the same Yardi database?
- Will everyone's system access be the same?

**Confirmation of training content and agendas**. Have the agendas been sent to the trainer for his or her review? Is everybody on the same page as to what the software training content will be? Will the trainees also be trained on what to do when they get back to their desk afterwards, and they are ready to go to work on Yardi Voyager? Will they have a checklist like the one included later in this chapter?

**Discussion of manual processes during the 2-3 day go-dark period**. How will incoming transactions be handled during the period in which the old system is down and the data is being converted in to Yardi Voyager? Should check receipts be held or deposited? What about invoices that come in? What about residential traffic? Should a side sheet be used? All of this needs to be discussed and documented, and it should be part of the communication strategy prior to the go-dark period. The examples at the end of this chapter are good templates.

**Post month of accounting balance-forward entries**. When tenant balances and open accounts payable balances are brought over from the source system into Yardi, they create a journal entry which will already be reflected in the general ledger activity for that particular month. So the effects of these balance forward entries must be reversed. The question is whether you want to book the tenant and open accounts payable entries in the current month or the next month. For example, if the conversion cutoff date is September 30th, would you want the entry to be in September or October? Most of our clients choose September because it keeps October, the first "live" month, "clean" of all of the in-and-out entries that occur when the conversion is occurring. Also, booking the entries in October can adversely affect your Gross Potential Rent (GPR) calculations if you use that reporting convention. The downside to booking in September is that it makes isolating the entries that need to be reversed a little more difficult, because there may be other entries already residing in the general ledger for September.

**Security.** Up to this point, I haven't talked about security–about what system rights users should have and what the general security strategy should be. We start this discussion at this T-minus six weeks with a planning meeting that includes all the members of the project team. The goal of this meeting is to define broad employee functions so that we can begin the process of assigning employees to a function. Examples of broad roles are:

- Accountants
- Accounts payable data entry
- Accounts payable check issuers

- Leasing agents
- On-site managers
- Regional vice presidents
- Super-user
- Reports only
- View only

Using Excel, create a matrix (yes, another one…) with the functions across the top as column headers and the system users as rows. Each employee is assigned to a security group. If you find that some employees don't quite fit into a security group, then it means that we need to create a new security group, on which may only hold one person. This is completely acceptable. It is better than the alternative of trying to fit this person into an existing group, thereby giving him or her either too much or too little system access.

The penultimate step is to determine what the menus should look like for each security group. Yardi Voyager gives you a lot of power to configure these menus. One old-school method you can use to complete this step is to print a screenshot of a complete menu and then use a highlighter to cross out the areas you don't want that particular security group to be able to access. Once you have the matrix and menu layout completed, we write a confirmation email to the project team. This email will request that the project team review the matrix and the screenshots and deliver any changes to the documents back to us. We usually give them a week to complete this review.

The last step is to go into Voyager Workstation Admin and use the available tools to set up the security matrix to which the project team has agreed. This should be tested by several people, by having them log in as one person from each security group. You will probably see some tweaks you need to make. Make the changes, and rinse and repeat until it looks perfect. During the go-live training, the trainees should log in to the system using their unique credentials. This is a good time to debug any problems prior to the go-live period. Know that it is not unusual for there to be some post go-live tweaking to a user security profile. After a few weeks, this should settle and you won't be changing it very often, if at all.

**Communication with HQ and site staff**. The next few pages illustrate a tactic one of our clients took after the pre-go-live meeting was held. They had a pre-kickoff meeting of their own in which the president of the organization stood up and said a few words about how this migration was a strategic initiative for them and how they needed everybody to be diligent and on-board for the upcoming changes.

The training director then created a series of checklists to be distributed to each of the 120 operating sites. These were passed out and monitored each week. (Notice how they made the on-site managers sign the check-list after all items were completed. This is a very savvy buy-in tactic.) Needless to say, this migration effort, which was larger than average in size, went very well. They were active partners in the planning process, both before we began and as the project progressed.

# YARDI "GO-LIVE" MEETING

Friday, September 2, 2005

11:00 AM

Conference Call

# Agenda

I.  A few words from our Prez...

II.  Roll Call

III.  Exciting times are upon us!

IV.  The Team

    a. IT Department

    b. Training Department

    c. Lupine Partners

    d. Yardi Account Managers/Programmers/Support Team

V.  The Plan

    a. Checklist Overview

    b. Envelopes and contents

VI.  Calendar

VII.  Q & A

VIII.  Hasta Luego, Muchachos!

## TWO WEEKS PRIOR TO ACCOUNTING MONTH-END (AME)

| PROPERTY INFORMATION | Property Name | Start date: |
|---|---|---|
| | Property Number | Manager: |
| **FIRST DAY** | 1. ☐ Confirm completion of lease file audit. (sent out by Drew) | |
| | 2. ☐ Assign "buddy" employee(s) to assist with completion of above task if not done. | |
| **GENERAL REVIEW AND** | 3. ☐ Review key reports for errors and contact help desk for corrections. | |
| **CLEANUP** | • F-106 | • 3.6.4 |
| | • 2.5.2 | • 3.6.6 |
| | • 2.5.4 pre-paid | • F-126 |
| | • 5.5 | • F-148 |
| | • Security deposit log | • F-119 |
| **ADMINISTRATIVE** | 4. ☐ Review general administrative procedures. | |
| **PROCEDURES** | • Make sure that: | • Remove all Wait List Persons and clear |
| | • certain security deposits match leases from file audit. | ledger balances. |
| | • delinquent balances are true balances | • Do your AME pre-close! |
| | • all completed work orders are marked as such | |
| | • ALL common accounts are at a zero balance | |
| | • lease expirations in RR match lease | |
| | • unit assets (new stoves, etc.) are on a spreadsheet if possible | |
| **ALTITUDES AND** | 5. ☐ Tell your staff how excited you are about YARDI! | |
| **ATTITUDES** | 6. ☐ Give them a chance to do a Q & A with you about what to expect. | |
| | • Multiple users can pay bills | • Information is LIVE |
| | • Multiple users can post money | • Reports can be retrieved from anywhere |
| | • Multiple users can enter traffic | with Internet access |
| | • Reports can be drilled down for more detail | • No more daily backups! |
| | • No more faxing ledgers for corrections | • No more posting NSFs |
| **ADDITIONAL** | 7. ☐ Forward questions you cannot answer to David Wolfe. | |
| **INFORMATION** | 8. ☐ Review job assignments and expectations. | |
| | 9. ☐ Review this task schedule and time frame. | |
| **ONE MORE THING...** | 10. ☐ Send pre-close reports to your accountant and to David Wolfe (another set of eyes looking for errors) | |
| | • 1.72 | • 1.7.5 |
| | • 2.5.4 (all) | • F119 |

ACKNOWLEDGMENT

(To be signed upon completion of all items)

Employee: _____ Date: _____

Manager: _____ Date: _____

Return original to Technical Training Department — Return copies to Manager and Employee

169

## DAY OF AME

| PROPERTY INFORMATION | Property Name | Start date: 9-19-05 |
|---|---|---|
| | Property Number | Manager: |
| **YARDI CONVERSION BEGINS…** | 11. ☐ Complete all pending activity in Rent Roll | |
| | 12. ☐ Make sure all money is banked and entered into Rent Roll by 3pm. | |
| | 13. ☐ Watch your email for any updates! | |
| | 14. ☐ Copy all monies received after close out and put in envelope marked No. 1. Deposit in bank as usual. | |
| **GENERAL REVIEW** | 15. ☐ Review key reports for errors and contact help desk for corrections. (This is the last chance to fix errors!) | |

| | |
|---|---|
| • F-106 | • 3.6.4 |
| • 2.5.2 | • 3.6.6 |
| • 2.5.4 pre-paid | • 1.7.2 |
| • 3.6.8 | • F-148 |
| • 1.7.7 | • F-119 |
| • 1.7.4 (close open batches) | • 1.7.5 |

| **ADMINISTRATIVE PROCEDURES** | 16. ☐ Refer to instructions on Rent Roll close-out procedures. | |
|---|---|---|

| | |
|---|---|
| • Verify accounting date | • Bulk bill late charges |
| • Rebuild indexes | • Print reports: |
| • Rebuild budget | 254,252,368,175,F119,172,177 |
| • Close open deposit batches | 2.5.4 (all,368,175,F119,172,177 |
| • Complete ALL pending activity | Balance reports |

| **WHAT WILL I DO FOR THE REST OF THE DAY??** | 17. Double- and triple-check your work |
|---|---|
| | 18. Send your five key reports to your accountant to review: |

- 172
- 174
- 175
- F119
- 254 (all)

| **ADDITIONAL INFORMATION** | 19. ☐ Start early and avoid the rush to get help. |
|---|---|
| | 20. ☐ Be ready for training beginning tomorrow. |

| **ONE MORE THING…** | 21. Are we excited yet? |
|---|---|

ACKNOWLEDGMENT

(To be signed upon completion of all items)

Employee: _____ Date: _____

Manager: _____ Date: _____

Return original to Technical Training Department — Return copies to Manager and Employee

**DOWN PERIOD...**
(DAYS DURING WHICH RENT ROLL ACCESS IS READ-ONLY AND YARDI IS ONLY USED FOR PAYABLE ENTRY)
TRAINING OCCURS DURING THIS PERIOD

| PROPERTY INFORMATION | Property Name: | Start date: |
|---|---|---|
| | Property Number: | Manager: |
| EVERY DAY | 22. ☐ Pull out the envelope with the corresponding date. | |
| | 23. ☐ Explain the use of the envelope and the importance of documentation while there are no computerized systems in place. | |
| GENERAL INSTRUCTIONS | 24. ☐ Place the following key reports in a central location so anyone can view or access them: | |

| General Instructions reports | |
|---|---|
| • F-106 (make sure updates are made to only one report to avoid duplicated rentals) | • 3.6.4 |
| • 2.5.2 | • 3.6.6 |
| • 2.5.4 pre-paid | • F-126 |
| • 5.5 | • F-148 |
| • Security deposit log | • F-119 |

| ADMINISTRATIVE | 25. ☐ Review general administrative procedures. | |
|---|---|---|
| PROCEDURES | • Make copies of all monies received that day and deposit as usual. | • Make copies of lease renewals and place copies in envelope |
| | • Document all new move-ins and move-outs. | • Place guest cards in daily envelope |
| | • Complete work orders on DW telephone service request form. | • Make sure unit assets (new stoves, etc.) are on a spreadsheet if possible. |
| | • Telephone answering forms | |
| ALL WORK AND NO PLAY... | 26. ☐ Keep good records so your manual updates are smooth! | |
| | 27. ☐ Use the proper envelopes (marked with the date). | |
| YOU WILL BE COMING TO TRAINING SOON! | • You can never have too much information. | • Less is NOT more at this time... |
| ADDITIONAL INFORMATION | 28. ☐ Get ready for your training class at a COMPUSA near you. | |
| | 29. ☐ Rent Roll will only be accessible as read-only. | |

ACKNOWLEDGMENT

Employee: _____ Date: _____

Manager: _____ Date: _____

Return original to Technical Training Department — Return copies to Manager and Employee

## GO LIVE!!!!

| PROPERTY INFORMATION | Property Name: | Start date: |
|---|---|---|
| | Property Number: | Manager: |
| **DAY 1** | 30. ☐ Add leasing people, as instructed in the "Navigation" chapter of your Yardi manual. | |
| | 31. ☐ Add Mode, Admin and Down Units, as instructed in the "Maintenance" chapter of your Yardi manual. | |
| **GENERAL INSTRUCTIONS** | 32. ☐ Open envelope no. 1, and begin entering your data from the dark period. | |
| | • Traffic – from guest cards <br> • Rentals- from applications <br> • Move-ins- from leases <br> • Move-outs- from walk sheets <br> • Lease renewals- from leases | • Cash Batches- from deposit slips <br> • Demographics <br> • Unit Inventory- from spreadsheet <br> • Enter work orders (all open and closed work orders from close-out) |
| **HELPDESK PROCEDURES** | 33. ☐ Refer to help desk procedures in Yardi Manual. | |
| | • Place call for help to the designated help desk number, and your call will be routed to the appropriate support person. <br> • Do not call your trainer directly; you may be calling while he or she is out of town training other people. | • Do not call your IT support technician directly, as he or she specializes in different products and may not be able to help you. |
| **LIFE AFTER TRAINING** | 34. ☐ Do not get frustrated. Sometimes it helps to walk away and take a breather. | |
| | 35. ☐ If you need additional training, contact your RPM to schedule. | |
| **ADDITIONAL INFORMATION** | You may be called upon for help from time to time. Do not be afraid to say "no" if you are too busy to help a fellow manager. Please refer the manager to the help desk. We are here to help. <br><br> **AND FINALLY, THANK YOU!!** | |

ACKNOWLEDGMENT

Employee: _____ Date: _____

Manager: _____ Date: _____

Return original to Technical Training Department — Return copies to Manager and Employee

What will Lupine Partners be doing during the go-dark process? On most projects, it's three things:

1. The project manager is acting as the master of ceremonies and the central communication point between you, the client, and Yardi, the software company.

2. Our data conversion professional is executing variable data queries and staying in touch with the appointed data validators.

3. Our training professional is executing the training program at the same time the variable data is being converted.

The go-dark planning meeting essentially compresses the project into a six-week period. You could almost throw away the original work plan at this point. (**But don't,** as it got us to this important milestone in the project!). At this point, you may not believe that you can actually get all of the work done in a mere six weeks. Others have stood where you do now. You can, and will, finish. It may feel like a "messy house" to you. I can tell you with 100% assurance that you WILL have a messy implementation house at this point. Our goal is to just get you moved in by the first day of go-live, not necessarily to have you living in a beautiful house that first day. We'll have some punch list items and that is usually fine. We just need to keep working the plan–albeit a new, slimmed-down, finely-honed six-week plan. At each status meeting we will check off more work tasks as they are completed and the funnel will get smaller and smaller, until you look up and realize that you have crossed over to the other side and "gone dark."

We'll only be in this apocalyptic phase for a day or two, during which time you may feel like Dorothy in "The Wizard of Oz" when she first steps out of her house after the tornado, from black-and-white ("dark") to full Technicolor ("live"). We're in the home stretch, folks. See you in Oz.

## *CHAPTER SUMMARY*

1.  *All of these work efforts to date have been choreographed into the execution of the final implementation effort, which is called "go-dark."*

2.  *The "go-dark" period is when you have turned off your existing source system and will operate without a system temporarily while your data is being converted from your source system into Yardi.*

3.  *The goal of the go-dark/go-live time period is to decrease the downtime in which you have no system, and to decrease the amount of time from when the end users get trained to when they begin using the new system.*

4.  *Coordinate and use manual systems and tasks during the go-dark process. Transactions still have to be tracked and cash still has to be deposited.*

5.  *Compile a list of first day go-live activities. These activities should be part of the end-user training that occurs during the data conversion/go-dark process.*

6.  *Have people other than the data conversion resources validate the data that has been converted.*

7.  *Compile a list of tasks going forward for the system users after they have gone live on the new system.*

8.  *Approximately 45 days before your scheduled go-live date, we will have a key strategy meeting with you where we will choreograph what the last 5-10 days before you go live will look like.*

9.  *The go-dark planning meeting compresses the project into a period of six weeks.*

# CHAPTER 12

# COMING OUT
# THE OTHER SIDE

At this point, congratulations are in order. You have made it through one of the toughest things you'll do as a professional–you've successfully migrated from one software system to another. You are now "live." Realize that some of your new users may be struggling as they begin to use the new system. Even though they just went through training, they may have been using the old system for years or even decades. People have different personalities, too. Some embrace these changes with a shrug of the shoulders while others may still be in denial that the "times they are a-changing."

There are several things that can be done to reduce some of the anxiety and fear after go-live.

One, have your support desk and/or hotline ready to field any incoming calls. Do not roll your eyes as the calls come in. People are afraid, and they may feel stupid having to make the call in the first place.

Two, have quick cards ready to go. Quick cards are one-page guides that tell users how to perform certain tasks within the software. These cards can reduce the number of help desk calls by 80% just by anticipating what the problems will be. Distribute the cards to the system users at the training they receive just before the go-live date.

Three, proactively check in with the users and ask them how things are going. Give them permission to call you–remind them that it's okay, and in fact, encouraged, to ask for help.

If you are receiving help desk calls into the third or fourth week following go-live, this means that the users were not fully trained. The most common offender (as discussed in Chapter 10) having gone cheap on the training and having jammed too many people into the training session. The old phrase, "You can pay now, or pay later" is true. You "pay" in that you are going to spend a lot of time/money reversing and correcting transactions made by people in the field who do not totally understand the software. And there is the internal cost of the despair and frustration of well-meaning people who cannot do their jobs because the training sessions were hijacked by overcrowding.

But if this isn't the case, and you are now live, what exactly does that mean?

1.  You are no longer processing transactions on the old software system.

2.  Your Yardi Voyager users have been trained on the proper use of the system, with a training session configured to each person's unique function in the organization.

3.  Your Yardi Voyager modules have been configured for your organization's unique needs.

4.  You have a beginning balance in your general ledger as of the go-live date. You may also have general ledger historical transactions, depending on decisions you made at the beginning of the project.

5.  You have entered or imported open invoice amounts by vendor and by property.

6.  Your static data has been entered or imported into the system.

7.  Your variable data has been imported into the system. All of the variable data has been validated and checked during the "dark" period that occurred the week before. You are secure in the knowledge that your tenant data is correct and that all relevant tenant balances and recurring rent charges are correct.

8. Your staff is going through the punch list items that were compiled during the few days they were out for training.

9. All operating reports, both canned and custom, have been tested and are on the proper user's menu.

10. Each user has a security profile that gives him or her rights to the Yardi Voyager functions he or she needs, but not to the functionality he or she doesn't need.

11. You are able to post next month's rent.

12. You are able to collect money and to apply that money to the correct tenant and for the correct charge code.

13. You are able to enter invoices.

14. You are able to write checks.

The last piece of the go-live effort will be to issue financial statements out of the new software system. This usually happens approximately 40 days after the go-live date; if your go-live date was October 1, your October financial statements would be distributed on or about November 10.

Once the financial statements go out, you should consider yourself to be successfully implemented. There are still some things to do, however. It may be a good time to fill in any data "holes." One area in which there may be holes is tenants. Depending on the system from which you are converting, you may have only gotten enough information out of your previous system to establish a tenant in Voyager for purposes of billing and collecting rent. But there may be additional data that you want in the system for that particular tenant. You can compile that data into a spreadsheet and then import it into Voyager.

Also, you may have some phase two projects. These projects are generally dependent on foundational data being in the system. Examples of phase two projects are cost recoveries, investment management, budgeting/forecasting, and bringing over the year-to-date vendor payments for purposes of processing 1099 payments by January 31. But know that for the most part you are done, and you can rest. You have gone into the "darkness" and come out the other side. That said, a few months from now you will realize that you are not using all of the (relevant) functionality of Yardi Voyager. In the next chapter, I will discuss our process to help you get more out of it.

If you have been on the project team, here's a note to you: You are now released from your obligation. You are probably tired and ready for a break. But you should also feel some pride. Correction: you should be full of pride! You've come a long way, baby. Go home, have a glass of wine (or two!), pet the dog and come back tomorrow ready to use this terrific software package called Yardi Voyager—the system you were so instrumental in bringing to fruition for your company. Good work.

## *CHAPTER SUMMARY*

1. *You made it through one of the toughest things you'll ever do as a professional–successfully migrating from one software system to another.*

2. *Employees have different personalities; some embrace these changes with a shrug of the shoulders, while others may still be in denial.*

3. *If you are receiving help desk calls into the third or fourth weeks after the go-live date, this means that the users were not fully trained.*

4. *Once the financial statements go out, you should consider yourself to be successfully implemented.*

5. *If you served on the project team, go home, have a glass of wine, pet the dog and come back tomorrow ready to begin work as a user of Yardi Voyager.*

CHAPTER 13

# GETTING MORE BETTER

Fast-forward from Chapter 12 (go-live) to a point six months in the future when you may be suffering from Post-Implementation Exhaustion Syndrome. On the one hand, you feel like you have it made in the shade, but on the other you sense somewhere in the back of your mind that there's more you should be doing with the software. But you can't seem to get the energy or the focus to either start the implementation project back up, or decide what you should be working on or where to begin. You may only have so much organizational bandwidth available to facilitate another major software initiative.

The problem is further magnified by the fact that Yardi continues to do a terrific job of adding functionality through product development and acquisition. This may lead you to suspect that there are aspects of the system that you could be utilizing to create efficiencies in your organization. You could be paying through your annual license fee for functionality which you're not actually using. To put it another way, you may not be optimizing your return on your annual investment.

If you are in this situation right now with regard to Yardi Voyager, you may also have:

- The perception that you are paying more for the software than the benefits you are receiving from it.

- The feeling that you're being left behind. (This may be particularly acute if you're returning from YASC or another conference at which you heard how other organizations are using the software.)

- The belief that you're still doing too many things manually even after using the software for a period of time.

- A feeling of resistance when you're contacted by well-meaning, got-your-back Yardi sales professionals because you don't know whether the additional functions and features they are presenting to you actually solve problems for your organization.

- A fear of being on the "bleeding edge" of technology–of being an early adapter, a guinea pig, going where no man has gone before...

## AN APPROACH

Here is how we have helped clients to work through this phase and get more out of the software:

First, we review your database for data incongruities and document the findings. This review is done offsite, before we even meet with anybody in your organization. One of the many complaints we've heard about the software is that the reports are sometimes "wrong." This is almost always a result of data issues, not programming ones. Some examples of residential data incongruity are:

- Past leases with no move-out date

- Prospect traffic with no source

- Concessions granted that do not show up in recurring charges (indicating possible fraud)

- Leases with no security deposits (indicating possible fraud)
- Tenant records with no SSN

Every real estate vertical, and every client, has their own version of what is congruent and what is not with regard to their own data. After the database review, a list of all possible data issues is presented for your consideration. Next, all relevant system users from all levels of the organization are interviewed individually and asked questions like:

1. How do you use the system on a daily, weekly, monthly, and yearly basis?

2. What is working for you and what is not? What slows you down?

3. What are your frustrations with the Yardi system?

4. Can you present the reports you use for both internal and external purposes?

5. What functionality do you think you should be using in Yardi, but aren't? Why?

The interviews are easy–they only last 30 to 90 minutes, and they're conducted in a safe, secure, and non-judgmental environment. An interesting dynamic that I've noticed over and over again that interviewees disclose things that they will never tell their employers. When we ask the right questions and use the proper techniques, the employees' unique truth surfaces. In order to maximize system usage, you need their candid input.

Third, we will combine the data incongruities, your disclosures, our observations, and our experience and knowledge to create a tailored plan with an emphasis on these 3 areas:

**Gaps.** An analysis detailing where there is a functional "gap" between your uses of the system and the available functionality will be performed. This gap analysis will take the current Yardi development road map into account.

**Recommendations.** The review findings will be documented and accompanied by recommendations for moving forward. The recommendations may or may not include the purchase of other Yardi offerings. Our finding and recommendation documents are brief, usually in the respectful range of 7 to 15 pages, and contain only hard-earned, direct advice to you.

**Strategy.** A road map telling you how to get from "here" to "there" will be created, one similar to the map created during the initial implementation. We usually list these plans so that the priorities are ranked as follows: Low-Hanging Fruit, Mandatory, High, and Medium. There are some initiatives that sometimes be accomplished in a morning and provide quick pain relief. Those items may appear at the top of the list; we like to knock those areas out, and then focus on the Mandatory initiatives and work our way down the list.

Four, you take action based on the road map created for you.

## YOUR INVISIBLE EMPLOYEE

Whether you realize it or not, Yardi Voyager is your Invisible Employee–available and ready to work seven days a week, 24 hours a day. The question for you to ask yourself is this: Even though Voyager (your employee) is a workhorse, are you keeping it busy enough to warrant the very high salary you pay it every year? Would you pay a human employee and accept only a 35% effort? You should be squeezing every bit of productivity that you can from your highest-paid employee. It will do whatever you ask–but you *do* have to ask.

View Yardi Voyager as a very narrow profit center. The revenues are not hard cash or accrual earnings, but rather benefits, both direct and indirect. Benefit examples are:

- Redirection of employees' time and focus
- Improved productivity
- Better decisions

- A possible decrease in FTEs

- Tactical rather than mechanical work

- Reassignment of human resources

- Elimination of errors

- Increased employee satisfaction

- Decreased employee turnover

- Scalability for growth

Here is a representative list of system functionalities identified for clients during this process:

| | |
|---|---|
| *Fraud Detection* | *Exception Reporting Strategy* |
| *Automated Distribution of Financial Statements* | *Delinquency Notes* |
| | *Property Attributes* |
| *Notifications* | *Market Surveys* |
| *Vendor Insurance Tracking* | *Report Inventories* |
| *Recoveries Setup & Strategy* | *Using Current Report Versions* |
| *Lease Abstract* | *Import/Export Strategies* |
| *GPR Reporting* | *Security* |
| *Tracking Variances* | *Analytics* |
| *Portfolio Level Reporting* | *Task Runner* |
| *Custom Cash Flow Statements* | *Attributes* |
| *Reporting on Percentage Owned* | *Report Scheduler* |
| *Attachments Strategy* | *MCA* |
| *Use of Memos* | *Performance Tables* |
| *Tracking Mortgages* | *Intercompany* |
| *Management Fees* | *Custom Tables* |
| *Account Trees* | *Allocations* |

You should view this post-implementation work as another implementation, one that uses a lighter version of the tools and tactics discussed in this book. Agree on scope, write a project plan, check in via status meetings, and test/orchestrate the implementation as best as possible. Write out the tasks for each project and sub-project. Don't just wing it. The same failure points that were discussed at the beginning of the point still apply here. You've come this far—why throw away the discipline that has worked so well for you?

One of the more confusing parts of migrating from one software platform to Yardi Voyager is the use of new terms, phrases, and words. Software vendors and consultants (like me) are guilty of using certain phrases and not explaining what they mean to our clients. In Chapter 14, I define and clarify many of these words and phrases for you.

## *CHAPTER SUMMARY*

*1. At this point, you may be suffering from Post-Implementation Exhaustion Syndrome.*

*2. You may only have so much organizational bandwidth available to institute another major software initiative.*

*3. You may suspect that there are aspects of the system that you could be utilizing to create efficiencies in your organization.*

*4. You may perceive that you are overpaying for the software when compared to the benefits you are receiving.*

*5. You may feel that, even after being on the software for a period of time, you are still doing too many things manually.*

*6. Whether you realize it or not, Yardi Voyager is your Invisible Employee–readily available to work seven days a week, 24 hours a day.*

*7. You should view this post-implementation work as another implementation, one that uses a lighter version of the tools and tactics discussed in this book. Agree on scope, write a project plan, check in via status meetings, and test/orchestrate the implementation as best as possible.*

# CHAPTER 14

# BIG WORDS

One of the more confusing parts of migrating from one software plat-form to another is the use of new terms, phrases, and words. Software vendors and consultants (like me...) are guilty of using certain phrases without explaining what they mean to our clients. The major offenders are listed below:

**Account Trees**. Account trees organize account information for finan-cial reports. The trees are a virtual remapping of your chart of accounts including headings, rows and totals. They can control the "row" format of a financial statement. Using account trees, you can generate reports with account numbering and ordering to fit special account requirements without having to create and maintain multiple charts of accounts.

**Analytic/Executive Reports**. This is a report processing and display environment that supports multiple columnar report types. Separate analytics reports exist for financial statements, journal registers, com-mercial tenants, residential tenants, retail tenants and budget & fore-casting. Analytics reports can also be added to the executive dashboard. They are included with the Voyager software. The primary differences between analytics reports and a report that outputs to the screen or to Crystal Reports are: 1) the environment has a static filter that stays at the top of the screen when you run your report, which allows you to easily modify the filter, and reprint the report; and 2) an analytics report typically has multiple report format (and even data output) choices for each analytics screen.

**Cash Accounts vs. Bank Accounts**. "Cash account" refers to the G/L cash account in your chart of accounts. "Bank account" refers to a specific bank account. You can have multiple bank accounts that link to the G/L cash account on your chart of accounts.

**Crystal Reports**. This is a report-writing and presentation environment. Reports that are created using Crystal Reports are generally presentation-quality reports with many formatting options, including the option to place data elements in something other than columnar format. There are exceptions, but generally these reports do not output well to Excel for editing, and do not have drill-down capability. You will probably want to use this environment for reports that need to be distributed in hard copy.

**Crystal Viewer**. This is an Internet Explorer add-in that allows you to view reports that were created in the Crystal Reports format on your screen.

**CSV File**. A "comma separated value" file, which looks very similar to an Excel file but doesn't retain any of the formatting that is used in Excel. These files are used to store the data that will be imported. The data fields are separated by a comma. (The commas can be viewed if the CSV file is viewed in a text editor like Notepad.) These files are used in conjunction with import scripts or FMT files.

**Custom**. The term "custom" is used in relation to Yardi when referring to reports or tables that are not core to the system. Custom tables or reports are those that are created/defined by the user.

**Custom Report**. A custom report is one that is either created or modified to fit your company's specific needs. Modifications of existing reports may include adding columns, changing titles, etc.

The other type of custom report is one that is created from scratch—designed by, and written for, your company.

**Custom Table**. A custom table (also called a one-to-many table or 1toM) in Yardi is one that is defined by the user to store additional information about a key identifier in the system (e.g., property, tenant, vendor). An example would be an insurance table attached to a property to track the various types of insurance carried by the property.

**DBO Login**. The Database Owner, or DBO, login is the highest-level (least restricted) access available in the Yardi system, as it bypasses all group and user security options. This login will also allow direct access to the underlying system database if you have access to the proper tools. Whoever serves as the system administrator for your Yardi installation will use this login to access the system initially while normal user logins are being created. It is not recommended that anyone perform day-to-day work using the DBO login because transactions are not tagged with a specific user name.

**Entity, Property and Accounting Entity**. Entity, property and accounting entity are used synonymously, and represent the level at which financial information is stored and reported. You will process all financial transactions at this level.

**ETL**. ETL stands for Extract, Transform and Load. ETL is a conversion tool within the system that uses Excel templates to simplify the data import process.

**Filter**. The filter captures the run-time parameters for records you are accessing or reports you are processing. A filter is used to limit the number and/or type of records returned by a report or function.

**FMT File**. A FMT file is used to import financial transactions (charges, receipts, journal entries) into Yardi through the "import transaction" function. It tells the Yardi system the order in which the data is stored in the CSV file.

**Import Script**. An import script is a text file that contains instructions for the system on where to import data. It contains the table name and the data field names that will be populated. The import script is used to import non-financial data.

**Import Trans**. Import Trans is a function within Yardi that allows you to import various transaction types (journal entries, invoices, checks, charges, receipts) into the system. The Import Trans function uses the FMT/CSV file combination or an XML file. This functionality is used often during the conversion period.

**Import Trial Balance**. Import trial balance has a similar functionality to that of Import Trans. Like the Import Trans function, the import trial balance function imports trial balance activity in the form of a journal entry. However, the import trial balance function also gives you the option of importing trial balance balances (rather than the activity) and then having the system calculate the month-to-date activity.

**Intercompany**. Intercompany functionality provides a means to track the amounts due to and/or from entities that have an agency relationship. There can be multiple intercompany relationships set up in the Yardi system, but each relationship is always tied to a specific, unique bank account. Each relationship can only have a single bank account owner, but there can be multiple entities that draw from, or contribute to, that bank account and owner. This functionality can be used while processing vendor checks, entering cash receipts, or making a general ledger journal entry. The most common scenario in which Intercompany is used is when you have an entity that pays invoices on behalf of one or more properties. Through the Intercompany functionality, you can easily keep track of what each property owes the paying entity.

**One-to-Many (1-to-many)**. One-to-Many (or "custom") tables are user-defined tables created to store additional information in the system about a property, tenant, vendor, etc.

**Package**. This is a program enhancement that either adds or removes functionality beyond what can be changed in the normal system setup menus. There are many packages included with the Yardi system, and they can be found on the System Admin menu on the Quick Menu under "Load Packages." You will find that other packages might also be available via your Yardi Account Manager, so if you desire extended functionality, you should always check with him or her before considering any custom programming. No package should be loaded onto a production (live) system before you completely understand the ramifications and have tested the new functionality on a test copy of the production database.

**Property Control**. You use the Property Control screen to define property defaults and retail parameters for a property. Some examples would be: property type (residential, commercial, etc.), the default cash account, and the late fee structure specific to that property.

**sCode**. You might hear people at Lupine and Yardi refer to an sCode. They are nothing more than IDs – like property ID, unit ID, etc. The sCode is the actual "behind the scenes" name for many system IDs. This is a unique code; two tenants cannot both have the same sCode. You may have two tenants named John Smith, but their sCodes will be different.

**Workflow**. You can create workflows to manage multi-task processes in recurring projects such as steps in the lease setup process, steps in the vendor setup process, etc. These workflows can be attached to multiple screens throughout Voyager to assist the user in managing the process without missing critical tasks. Workflows are like to-do lists, and there are three types of workflows that you can set up in the system: punch-list, punch-list sequential and user-directed.

A punch-list workflow will display all the tasks on the workflow (to-do list) and the user can perform these tasks in any order. A punch-list sequential workflow forces the user to complete the tasks on the workflow in a specific order. The user will not be able to move on to the next task without completing the previous one. A user-directed workflow allows the user some choice in how to proceed through it. For each task, there are one or more tasks that can be performed, depending on how you set up the workflow.

**yCheck**. This is a stand-alone program that is installed on a user's computer to allow him or her to process vendor payments via checks. Only computers with this program installed can be used to process checks. This, combined with Yardi group security, helps ensure that only the users that you designate can process checks.

**yEmail**. This works in conjunction with Yardi Conductor; it's the program used to send report files by email to designated recipients.

**YSL**. YSL stands for Yardi Spreadsheet Link. This is a reporting function in Yardi that allows you to export financial and non-financial totals from your database into a pre-formatted, user-designed Excel worksheet or worksheets to create custom reports in Excel.

CHAPTER 15

# FREQUENTLY ASKED QUESTIONS

**What is the bare minimum that has to be done from my side in order to get the implementation started?**

Surprisingly little. You don't necessarily need to have the contract with Yardi signed to begin the implementation process. I would say the two items on the short list would be: 1) the creation of a project team and 2) the discovery process (four to six hours).

Three days after the initial discovery process, we will conduct a kickoff meeting at which we present the project approach to you. Your only obligation is to show up and participate. From there, we'll go into the module design and configuration meetings. There is a lot of work for you to do during these meetings, but you still don't technically need to have the software loaded and available to you, because we are going to use sample data on our laptop as we go through the module design and configuration meetings. After that, though, the project may stall if you haven't signed the contract with Yardi and had your database established. This is because the next step in the process is to begin actually doing the steps that were decided upon at the design and configuration meetings.

**What is the typical number of consultants you employ on an implementation engagement?**

It's usually one consultant, but can be two if the engagement is larger or if certain subject matter specialties are required. The reason we can keep it so low is that we're good at project management *and* in the subject matter itself, and can fill a number of roles. All the consultants at Lupine are adept and trained in both classic project management techniques and on most, if not all, of the modules within Yardi.

**What are your guarantees on a software implementation project?**

There are several things that we do guarantee:

1. That you will be led.
2. That we will work hand-in-hand with Yardi to be one cohesive unit.
3. That we will have the required product knowledge.
4. That we have a proven methodology for implementing software.

The one thing that we can't guarantee is what your work effort is going to be, which is crucial to success. Lupine and Yardi can do everything correctly, but if our mutual client is not available to work on the project in the given timeframes, we can't guarantee that the implementation will actually occur at your desired go-live date.

**What is your process for estimating consulting costs on a Yardi implementation engagement?**

Since we have a proven implementation method that we use over and over, we do our estimating based on each segment of that implementation methodology. We have worked with over 100 Yardi clients over the years and therefore have a great deal of data to compare. We have accumulated metrics on how long certain tasks within the implementation are going to take, and have built an estimating matrix based on the size of different organizations. So after asking some key questions, mainly about the amount of history that the client wants converted and whether the client or Lupine is going to do data conversion), we are able to provide a consulting estimate that historically has been on the money. The reason we are always so close to budget is because our methodology has solidified and doesn't change. We implement every client the same way (methodology-wise) and we implement every client differently (because you all have different needs and issues).

### How do you stay current on Yardi's product offerings?

It's not easy, because Yardi is creating new products all the time as well as enhancing existing ones. Several things that we do to stay current are:

1. We attend Yardi's annual consultants' conference at which they introduce and explain their newest product offerings.

2. We have access to their knowledge base on Clients Central, and use this repository frequently.

3. We have strong relationships with appropriate Yardi personnel, relationships that have been developed in the trenches as we've worked through difficult issues. (We're a band of implementation brothers, if you will.)

4. Working through problems during implementations; as certain issues come up, we handle them with the help of the Yardi support personnel. In the process, we learn more about all of Yardi's offerings.

5. Lupine has internal quarterly consultant meetings where we share knowledge of functionality our individuals have learned throughout the quarter. Each consultant is charged with creating videos of newly-learned functionality. These videos are put in an internal electronic library to be "checked out" as needed. Our clients can also check out videos to supplement their training.

## If we hire Lupine to lead our implementation, what will Yardi's role be?

If Lupine is leading the engagement, then Yardi's main job is to fix any problems in the software that may arise during the implementation. Sometimes they also find resources within their company with which to solve a problem. They work with Lupine, hand-in-hand, to resolve issues as they arise during the engagement. Yardi has a seat at the project team table, attends all of the status meetings, and is a fully functioning member of the implementation process.

## How many Lupine employees actually went through Yardi implementations prior to joining Lupine? (How many of your employees have sat where we sit right now?)

It's a good question, and the answer differentiates Lupine from other consulting firms Angela Chaney, Debbie Graves, and Amy Beesley) went through Yardi implementations prior to joining Lupine. The experience of sitting on the other side of the table has proven invaluable to them in their careers at Lupine. We've been where you are now and we conduct ourselves accordingly, understanding that there may be some fear associated with the change in software packages. We never underestimate or minimize the emotions that sometimes accompany this process.

## What is your philosophy regarding the timing of end-user training?

We have established a methodology in which, in the two days between shutting down your existing systems and going live on Yardi, your user base is trained. The training of end users is simultaneous with the final conversion of data onto Yardi Voyager. This way, when the users get back to their desks after the training, they are live with Voyager. All of your properties, units, tenants, and recurring charges, along with tenants' balances at the time of the conversion, have been loaded onto Yardi and your staff can begin working in a live environment. This reduces the amount of time in which knowledge gained during the training can be forgotten.

## Do we have to have customized reports?

Absolutely not. The best bang for your buck is to come up with a strategy to use the canned existing reports in Yardi. Sometimes putting two reports together or running two reports will negate the need to actually create a custom report. However, if you desire a report that doesn't exist in Yardi, then there are resources (Lupine, Yardi, and other outside consultants) available to you. These people have written many custom reports and have a library of knowledge with regard to creating them.

## Who is responsible for converting the data from our existing system? Is this a service you provide?

It's really the client company's call to decide who converts the data. There are several options for you to consider. One is to hire Yardi's data conversion team to do the conversion; they do a terrific job. Another option is to hire Lupine to covert the data—this is a particular area of expertise for us. The third method is the one we're seeing more and more often, one in which Lupine trains you how to extract the data from your source system, and then how to import it into Yardi. Increasingly, clients are going this route, and in our opinion, it's the best value for your data conversion investment.

## Which real estate vertical markets have you performed Yardi implementations for?

We have provided services to the following vertical markets:

- Commercial
- Retail
- Residential
- Industrial
- Asset Management
- Fund Advisor
- Institutional Investment

- Public Housing
- Affordable Housing

## What is the delineation of the responsibilities between Lupine, Yardi, and the client during an implementation?

If Lupine has been hired to be project manager, our job is to provide leadership, subject matter expertise, and both module training and end-user training. Yardi's job is to make the software run as advertised. You, the client, are responsible for making operational decisions, for getting the modules set up correctly after we facilitate the module design and configuration meetings, and to make sure that all of the people who will be using the system receive training. The conversion of data will be assigned to one of the three organizations after the discovery meeting.

## What is the purpose of the discovery meeting?

The discovery process has two main goals. The first is to get organizational agreement on the project scope. It is rare that the entire company is in sync with what is going to be implemented and what the conversion process is going to be. Lupine can't manage a moving target. Therefore, we ask a lot of questions about scope. The second goal of the discovery process is to create custom implementation materials for our client to be presented during the implementation kickoff meeting.

The goal of the discovery process is **not** to begin making decisions regarding how the chart of accounts is going to be set up, what the unit type numbering methodology will be, or what the financial statements should look like, and so on. All of that will be handled in the module design and configuration meetings.

## How do you communicate project status?

Every week we send out a status report. For consistency purposes, we send it out on the same day of each week and around the same time. It is sent to the project team, other relevant people in your organization, and relevant people at Yardi Systems including the sales professional who sold you Voyager. The simple, but effective, format looks like this:

- a paragraph that gives a summary overview of the project
- a listing of tasks that were completed during the week
- a listing of tasks assigned but *not* completed
- a listing of key decisions made during the week
- a listing of tasks assigned for next week

Additionally, we use the status report as the agenda for the weekly status meeting we hold every week until we've gone live and the project team disbands. These status meetings are held by phone and are attended by personnel from Lupine, Yardi, and the client.

**How do we know that Yardi will support Lupine Partners as we all go through the implementation together?**

The answer is based on pure history. We have worked on many implementations with Yardi, and been hand-in-hand, side-by-side for all of them. We have an office chock full of testimonials from clients praising both Lupine and Yardi. Lupine and Yardi share a commitment both to the client and to one another's organization. Lupine helps Yardi, Yardi helps Lupine. It's a respectful and intentional relationship that we have built with one another.

**Have you ever had a project fail that required you and/or the client to walk away from the implementation?**

We've had this happen only twice in our 20-year history. In both instances, its foundation was in the conversion of the general ledger history. In both cases, the projects moved more quickly than the client's ability to extract, import, and validate this history. They were trying to convert too much history, and at the various go-live dates, they were always still several years behind, and wouldn't take the remedial steps necessary to get this history loaded. If you are bringing over general ledger history, you need to make sure that the time/money cost of the conversion is worth the benefit of having the history in Yardi.

**What, in your opinion, is the most important part of the implementation?**

Discovery, kick-off, module design and configuration meetings, weekly status meetings, training, and orchestration of the go-live process–these really are equally important, and they tend to build off of each other. However, if forced to pick one, I would say discovery, the process of getting everybody onto the same page. Discovery serves as the foundation for communicating an implementation approach. It's key. If it's done badly, the results will resonate throughout the entire implementation process.

# CHAPTER 16

# TALES FROM THE ROAD

Do you remember the Johnny Cash song with the lines, *"I've been everywhere, man. I've been everywhere...?"* At this point in my professional life, I sometimes feel that way. In this chapter, I will tell you some true software implementation stories that may be instructive for as you start your software journey. Good luck to you.

---

In 2010, I attended a software conference on the West Coast where I ran into one of my former clients. She was the Director of Information Technology for a company on the East Coast. In 2005, we were hired to implement her firm to Yardi Voyager. It was a difficult engagement–project team members were hired and fired in the middle of the engagement. As a result of this turmoil, implementation decisions were revisited, and in some instances, the decisions were changed which resulted in go-live dates being pushed back.

As we got to the end of implementation, nerves began to fray. This IT Director was a very nice woman, but as time went on, she became increasingly aggravated. She wasn't necessarily frustrated with me, but she was the corporate punching bag and she needed a venting outlet. And, apparently, that was me. She began getting tense with me in our daily and weekly conversations. But we got through the engagement, and we moved on.

When I saw her at the conference, we were catching up and she asked me how things were going at Lupine. I went through the good, the bad, and the ugly with her--and I must have mentioned some project in which the client project team members were struggling. She then asked me,

"Do people ever get upset with you on these types of consulting jobs?"

I almost spit out my drink. She saw my expression and the big grin on my face. "What?" she asked. I told her that she had "ripped me a new one" on a fairly steady basis over a period of two months. "I did?" she asked me incredulously. I could tell she had absolutely no recollection of being verbally abusive. All she remembered was that her company had been successfully implemented on Yardi. She loves her baby–all the labor pains and the difficult delivery are forgotten.

———

In 2001, I was involved in a major and difficult software implementation project. The deadline was tight and the data conversion obstacles were numerous. It was a big job. We were rolling out these properties in six groups, each with about ten properties in it. The rollout of the first group went well. Organizational stress was under control, and my client could now see that we were going to make it. Maybe we lowered our defenses.

For the second group, we went through all of the steps outlined in this book. Except, as it turned out, we skipped one. My client had hired "touchers" to go out into the field and sit with the on-site personnel as they went live. (This is not a bad strategy, by the way—if you can afford it.). On the go-live day, I received a call from one of the touchers saying, "David, we have a problem…"

Apparently, the cutover balances for each tenant were incorrect. But what was even stranger was that the balance was off by one tenant. In other words, the tenant in unit 101 had the cutover balance for the tenant in unit 102. The tenant in unit 102 had the cutover balance for the tenant in unit 103. And so on. Alarm bells went off in my head. I was pretty sure I knew what had happened, and it really pissed me off. I went and grabbed the source balancing reports in which the data was validated. And my worst fears were confirmed. If you have ever worked in Microsoft Excel and are dealing with a lot of data, then you can have "row slippage," in which certain columns "slip" by a row. In this situation, the validator was checking the totals, not the balances of each individual tenant. I went and found the offending party and asked her about the level of detail in her validation review. She cowered and hedged. I then showed her the tick marks where she had checked off the totals without looking at the individual detail. Long story short: I fired her on the spot and then went and told my client what happened. (Incidentally, this client now works with me at Lupine Partners.) Longer story short: you CANNOT skip this step. (We solved the problem by backing out the offending transactions, correcting the source spreadsheets by one row, and then re-importing. It took about 45 minutes.)

———

Same year, same client, same group 2…

We were importing tenants for one of the properties when I received a call. ("David, we have a problem…") It seems that the tenants in Group 1 were now getting transactions from Group 2. I stopped the presses, putting the entire conversion on hold including the properties that were already live and processing transactions. We took the system down so we could find out what was going on and keep a bad situation from getting worse.

The problem had to do with how we were assigning tenant IDs to the Group 2 tenants being converted. While leaving a gap between Group 1 and Group 2 tenants, we didn't think about the fact that these were college student-occupied properties. In student housing, there's something called "The Turn," which is the mass move-out and move-in of student tenants . In this case, our Group 2 imports butted into, and overwrote, the new move-ins for Group 1. We had a mess.

We were able to find out when our import was done and to restore the system to the stage just before we did the import—in essence, we performed a massive "NEVER MIND." We were able to turn back the clock and act like it never happened in the first place. I will never forget the feeling of looking at Brian (Wood ,who's now with Lupine…) when we realized what happened. That sick feeling in my gut. And the euphoria when we came up with the solution…

———

I have begun noticing an interesting phenomenon that first came to my attention several years ago: Increasingly, people do not want to schedule meetings to talk through the details of their software implementation. At Lupine, we do everything by appointment. Otherwise we would never be able to complete work for our clients. We would always just be handling incoming. LIFO, if you will, for the accountants out there…

These attempts at mature business dialogue usually occur before we are even officially engaged. We go back and forth for weeks just trying to book a phone meeting. The client asks me to call him or her at any time. I don't find that talking to a potential customer while he is at the drugstore or she is at home results in a productive work session. Even more curious is people's complete inability or desire to commit to a meeting time. They are kind of like the people who will not RSVP to a party, waiting until the last minute to see if something better comes along. As for me, I want to have productive discussions and I want to give them my complete and undivided attention.

In each and every situation like this, where I have had a flighty person as my point of contact, the implementations have been more difficult than they needed to be. Every single time. As you now know, these implementations are one big choreographed effort involving planning, intentional action, and systematic thought. The consulting equivalent of "just call my cell" does not work.

I no longer accept consulting engagements if the prospect will not agree on a day and time to meet. Frankly, it just isn't worth it. They're curious as to why I will not accept them as a client. Their behavior is an absolute guarantee of implementation failure, difficulty, and despair.

———

The most frequent and difficult discussion we have during the discovery phase of the project surrounds how much general ledger history to bring over. This was the case in a profound way for a major mall in Dallas that hired Lupine in the late 1990s to serve as the project manager for their software migration project.

The problem is not the difficulty of the task. Most of the time, extracting the data from the source system and then importing it into the new system is straightforward. The difficulty lies in the validation of the data. For example, let's say you are implementing 50 properties and you want to bring over five years of historical general ledger information. Here's the math:

50 properties x 5 years x 12 months = 3,000 trial balances to be validated

That's a lot of work. That's a lot of eyelid-drooping work. To compound the problem, what if the property IDs change, as they sometimes do? And what if the chart of accounts changes, as it almost always does? Well, now the job has gotten much, much harder because the chart of accounts is different. You can no longer just do an eyeball check of the data.

The Dallas client ended up walking away from the implementation just because of the general ledger history issue. Here is what happened: First, they insisted on ten years of general ledger history. Okay, that's bad, but not fatal. What *was* fatal was that nobody on staff would do the export of data from the source system. Nobody agreed to import the data into the new system. Nobody agreed to validate the data. And, to make the "perfect storm," the client would not pay an outside firm to do the work for them.

Yes, you are reading this correctly. They wanted the transactional history, but wouldn't do the work or pay for it to be done. Time moved more quickly than the client's ability to bring the information into the new system. Eventually, the project team disbanded and we moved on down the road. All of this carnage occurred because of this one notion around financial history. They would have been much better off brining over no general ledger history and just moving ahead with the implementation.

What a waste.

———————

Another, more recent, example of a company making a poor historical transaction conversion decision involved a client in the Midwest. This company wanted to bring over vendor payments made over a five-year period. This is not a problem at all; in fact, it's easier than bringing over general ledger transactions, as no validation is required. You simply import the transactions as checks and then make a reversing entry in the general ledger to offset the financial statement effect of the import.

But, no...

They wanted to do the import as described above, but did NOT want to reverse out the effects to the general ledger. They then wanted the historical general ledger transaction import to take into account the accounts payable import, essentially booking a delta or differential entry. It was really and truly a horrible idea. We objected to their decision in person, over the phone, and in writing. As loving parents to our client children, we tried to take away the keys to prohibit them from driving recklessly, all to no avail. They wanted to do it their way and it completely blew up on them. There were horrible after-effects and recriminations. And it just didn't need to happen.

Here's the lesson: One, listen to your consultant. Two, determine what you are really trying to do, and do a cost/benefit analysis, taking into account time, difficulty, and perceived upside. And, three, listen to your consultant...

———————

More and more often, we see implementation projects get behind right out of the box due to a client's inability to craft a chart of accounts. This wasn't always the case–it has only reared its ugly head in the last couple of years. And it still catches all of us at Lupine off-guard.

If you are struggling with creating a chart of accounts, here are some things to consider:

- If you are changing the chart of accounts, you can do this work prior to the implementation. In most (but not all) cases, the target software system does not have an effect on the structure of your chart of accounts. This work can, and should, be completed prior to implementation. You can pay now or pay later. Pay now.

- Don't overthink it. You can always add accounts later. Despite what other consultants may tell you, there really is not a best practices in this area, except that most charts go in this order: assets, liabilities, equity, revenue, and expenses. The number of characters is up for discussion, as is the numbering scheme. There is no right or wrong. My only rule is not to back yourself into a corner with your numbering methodology. Leave room for expansion. One extra digit will usually do the trick.

- At a minimum, you need to initially load the accounts to which you'll be mapping as a result of the general ledger history conversion. As I said above, you can add accounts later.

The inability to create a chart of accounts in a timely manner is the No. 2 reason why go-live dates get pushed back. (Number 1 is the conversion of too much general ledger history.)

---

In 1995, a client of mine in Santa Fe, New Mexico flew me in for the day to train a new employee how to use their property management software. Prior to the training, I asked the client several times if there was any preparation I needed to do prior to the one-on-one session. Each time, I was told just to show up for a freewheeling session. Against my better judgment, I said "okay." John picked me up at the Albuquerque airport and we drove about an hour to Santa Fe. I asked a few more times how he saw the day going.

When we reached his office, he showed me the conference room door and told me we'd be working in there. I walked into the conference room and there were 12 people sitting there, all expecting to be trained by me! I had no agenda, no computer, no computer setup—and to make matters worse, the topic was not one of my strongest.

I began to feel the sweat on my forehead and back, and to smell the stench of embarrassment and failure. I went to find John and did the 1995 version of WTF. He thought he had told me there would be an entire group to be trained. (We were JUST in the car together when I asked him that exact question!)

To make a long story short, it was a tough day but I got through it. Having a sense of humor helped. The big lesson learned is that you always assume that there will be more people attending than you anticipate. Always assume that the room will not be properly set up. Always assume that the technology will not work. Be ready for all contingencies, because when you are in front of a group of people, forces always seem to conspire against you. This experience made me a better consultant, because after I had to struggle against so many obstacles that day, I made sure that nothing like that ever happened to me again.

And it hasn't.

*Of travel I've had my share, man. I've been everywhere...*

# ACKNOWLEDGEMENTS

This is my second book. As Hillary Rodham Clinton wrote, it takes a village...

I would like to thank my clients. I feel very strongly that I was born to do this work–but I couldn't do it if you didn't hire my company to help you. One of the nice things about being a consultant is that you get better every day, week, month, year, and decade. You are always making mistakes, learning from them, and then passing the hard-earned knowledge on to your client constituents. There is an intimacy that builds over time. Of mutual trust between consultant and client. I don't have it with all of my clients, but I have it with many of them. You know who you are.

Next up is Jack Dicks and Nick Nanton from the Dicks Nanton Agency. I was already in the process of writing the book when they both convinced me to go a slightly different direction. They were right. I was wrong. And I'm better off for it. And thanks to Angie Swenson, agent to the stars, for pulling everything together for me. And hats off to Editor Emily Griffin for taking my scribbles and making them sing.

I am very proud of my 12-year association with Yardi Systems. It has been amazing to be on the outside looking in and to see the explosive growth this fabulous company has undergone. I am, and will continue to be, in awe of what you have built. I am indebted to Anant and Gordon for allowing my company to float around on the perimeter of your walls. I am especially indebted to John Pendergast and Terri Dowen for taking in Lupine at the beginning and working with me to foster a rather terrific working relationship. Thank you.

I'm not done. I work with a lot of people at Yardi Systems and it is impossible to list all of the people who have influenced this book. I am truly blessed to have working relationships with some extraordinary software sales professionals. I am grateful for your trust, guidance, honesty, and professionalism. To name a few: Stacy Blanchard, David Brown, Scot Cowger, Wendy Fitzpatrick, Alan James, Sam LaNasa, Neille Kommer, Tom McDermott, Daryl Pitts, Mike Pollasky, Spencer Stewart, Heidi Stilwell, Jennifer Swanson, and Michael Tuer. Thank you.

Yardi Systems is unique in that it actually supports the growing network of independent consultants who consult on Yardi's suite of products. And they really do support us. It's quite an amazing thing. To Becky Sanvictores, Nancy Bogg, and Lina Castanon–big hugs.

My stepdaughter, Emily Martin, started this project off by agreeing to transcribe all of my audio and video files. This is the office equivalent of having to dig a ditch. She did it with a smile on her face and nary a complaint. Good job, Snooks.

All of the employees at Lupine Partners live and breathe this methodology. They could have just as easily written this book. They are my traveling, wandering, nomadic band of all-stars. I wake up every morning knowing that the company is represented by the best and that they have my back. In order of tenure:

Amy McNeill has been my trusted lieutenant for 14 years now. She is a good friend and a trusted adviser. Her answer to any request of mine has always been "Yes!" She has also stood by me during periods of personal turmoil.

Brian Wood was the IT director when I ran my first Yardi implementation way back in 2001. As I told him back then, he made me raise my consulting game due to his high standards. He is now in his tenth year at

Lupine. He is a monster talent and is a valuable resource to everybody–Lupine and clients alike.

Maggie Wolfe, aka "Puppy," is my daughter. She is her sixth year with the company and has become my "right-hand man," as they say. We are VERY much alike and are able to communicate in a special Wolfe language. (We have also determined that every business situation can be found in an episode of "Seinfeld"!) I know that it ain't easy being the boss' daughter…

Angela Chaney, in her fifth year, also worked with Brian and me on that first Yardi implementation back in 2001. Among her many strong traits, her ability to learn from her mistakes is quite remarkable. She's applies the rule "one and done" when receiving input and correcting herself. Also, if you are ever driving through a blinding snowstorm in Colorado, she is your gal…

Debbie Graves is winding up her first year at Lupine, and she too worked with Brian, Angela, and me on that first implementation long ago. We go way back. Debbie's forte is training and she has filled a void that existed at Lupine for a long time. She is quick-smart, and has a passion for teaching others how to get the most out the software.

Amy Beesley started her employment with us about four weeks before the completion of this book. Nonetheless, she played in a major role in its completion. She reviewed the first draft of the book and made substantive and grammatical changes and suggestions—most of which I incorporated into the final draft. She is a most welcome addition to our consulting family.

I'll end with this. Most of the credit in writing this book goes to my wife Susan. I have made a lot of mistakes in my life. I was forced into a difficult bankruptcy as a very young man, and I have suffered through betrayals, humiliations, and despair. And out of the ashes, through the serpentine path of life, came Susan Culpepper. We married six months after meeting. It just took one dinner and one hockey game together. Game, set, and match. The manner in which she lives her life and the support she gives allows me to operate in "wild man" mode. I am most definitely not an easy man to live with. Hopelessly self-focused and pre-occupied. But she pulls it all off (seemingly) effortlessly. (Wine doesn't hurt…) Thank you, dear woman, for saving me. Life is good. Or as Dan Patrick says on his daily radio show, "Every day is the Super Bowl."

David Wolfe
Dallas, Texas
June 30, 2012

# PROTECTING YOUR INTERESTS!

You are embarking on an interesting professional journey. Selecting, designing, and implementing new software, while rewarding, is not without its challenges. Our firm has been providing these protective services since February 1993. During this time, we have successfully provided software consulting solutions to hundreds of clients. While we learn something new with each client, there is very little ground that we haven't covered before. At this point, most of our education is in the nuances and margins.

You can find additional resources to help you through this challenging process on our website, www.lupinepartners.com. There, you'll be able to sign up for our software email series, watch videos of me discussing various topics of interest, and access our library of consultant videos and tips. All of these resources and more are available to you on the website.

 I look forward to hearing from you about your unique set of issues. I can be reached at:

dwolfe@lupinepartners.com
214.953.1032

Good luck to you!

www.ingramcontent.com/pod-product-compliance
Lightning Source LLC
LaVergne TN
LVHW011941060326
832903LV00051B/295/J